Praise for *Power of the People*

" … a clearheaded, plainspoken primer on the nation's looming energy crisis that deftly and authoritatively identifies the issues and points to possible solutions. A copy should be on the bedside table of every presidential candidate, policy maker, and business owner—and just about everyone else who cares about the future of this nation."

—Alex Taylor, senior editor of *Fortune* magazine

"*Power of the People* is a terrific energy primer for the layperson and expert alike. An easy and highly informative read!"

—Heidi VanGenderen, energy advisor
to Colorado governor Bill Ritter

"*Power of the People*'s vision for 'the now' is engaging and enlightening. With Tombari's straightforward facts in hand, everyday consumers can make intuitive energy choices that will better all of our lives."

—David Terry, executive director, Association of State Energy
Research and Technology Transfer Institutions

" … a crystal clear handbook on what to do from an expert who has dedicated her career to the issue of clean energy."

—William Becker, executive director,
Presidential Climate Action Project

" … provides readers with an entertaining insider's view of America's energy revolution."

—Chuck Kutscher, past chair,
American Solar Energy Society

"For decades, Carol Tombari has been deeply involved in the discussions and debates about energy efficiency and renewable energy. … In *Power of the People* she balances the good and the bad to give us the background and the information to understand the technical, economic, and public policy issues that will drive the development of clean energy for this early part of the twenty-first century."

—Frank M. Stewart, president, American
Association of Blacks in Energy

SPEAKER'S CORNER is a provocative series designed to stimulate, educate, and foster discussion on significant topics facing society. Written by experts in a variety of fields, these brief and engaging books should be read by anyone interested in the trends and issues that shape our world.

More thought-provoking titles
in the Speaker's Corner series

Beyond Cowboy Politics
Colorado and the New West
Adam Schrager, Sam Scinta, and Shannon Hassan, editors

The Brave New World of Health Care
What Every American Needs to Know about the Impending Health Care Crisis
Richard D. Lamm

Condition Critical
A New Moral Vision for Health Care
Richard D. Lamm and Robert H. Blank

Daddy On Board
Parenting Roles for the 21st Century
Dottie Lamm

The Enduring Wilderness
Protecting Our Natural Heritage through the Wilderness Act
Doug Scott

Ethics for a Finite World
An Essay Concerning a Sustainable Future
Herschel Elliott

God and Caesar in America
An Essay on Religion and Politics
Gary Hart

No Higher Calling, No Greater Responsibility
A Prosecutor Makes His Case
John W. Suthers

One Nation Under Guns
An Essay on an American Epidemic
Arnold Grossman

On the Clean Road Again
Biodiesel and the Future of the Family Farm
Willie Nelson

Parting Shots from My Brittle Bow
Reflections on American Politics and Life
Eugene J. McCarthy

Social Security and the Golden Age
An Essay on the New American Demographic
George McGovern

Stop Global Warming
The Solution Is You!
Laurie David

TABOR and Direct Democracy
An Essay on the End of the Republic
Bradley J. Young

Think for Yourself!
An Essay on Cutting through the Babble, the Bias, and the Hype
Steve Hindes

Two Wands, One Nation
An Essay on Race and Community in America
Richard D. Lamm

Under the Eagle's Wing
A National Security Strategy of the United States for 2009
Gary Hart

A Vision for 2012
Planning for Extraordinary Change
John L. Petersen

Power of the People

America's New Electricity Choices

Power of the People

America's New Electricity Choices

Carol Sue Tombari

FULCRUM
GOLDEN, COLORADO

Library of Congress Cataloging-in-Publication Data

Tombari, Carol.

Power of the people : America's new electricity choices / Carol Sue Tombari.

p. cm. -- (Speaker's corner books)

Includes bibliographical references.

ISBN-13: 978-1-55591-626-8 (alk. paper) 1. Renewable energy sources. 2. Electric power production--History. I. Title.

TJ808.T66 2008

333.793'20973--dc22

2007039832

Printed in Canada by Friesens Corp.

0 9 8 7 6 5 4 3 2 1

Editorial: Haley Berry, Katie Raymond

Design: Jack Lenzo

Fulcrum Publishing

4690 Table Mountain Drive, Suite 100

Golden, CO 80403

800-992-2908 • 303-277-1623

www.fulcrumbooks.com

I dedicate this book to the future generations, especially to my children and grandchildren—Chris, Tami, Colin, Ron, Julie, Jackson, Emily, Mason, Logan, and others yet to come. They and their children will inherit the earth—or what's left of it after we refashion both the land and the atmosphere in our grab 'n' go attitude toward depleting resources as we feed our insatiable appetite for cheap energy.

Although this book is dedicated to the generations to come, I am indebted to those who have gone before.

To my grandfather Bob Mac, a lifelong fisherman and licensed guide, who told me—years before we read it in the popular press—"Lake George is dying." Sure enough, years later we learned that power plant emissions from the Midwest were injected into the jet stream and concocted an acid rain that, when spilled on this pristine lake in New York's Adirondack Mountains, smothered the resident trout.

To my mother, Elizabeth MacWilliams Tarbox, who taught me how to write and made me want to.

To my dad, Tom Tarbox, who instilled in me the passion for cool technology. His was the air-cushioned landing gear. (There, ol' Tom. I got it in a book!)

Contents

Part II: Where We Need to Go and How to Get There

Preface

Noted author and columnist Thomas L. Friedman was among the first prominent voices to articulate the grave and wide-ranging consequences of the United States' pitiful dependence on oil imported from the Middle East. In recent years, he has called—over and over again—for a national commitment to develop our domestic alternative and renewable energy resources so that we can start to back off foreign oil. Friedman compares such an effort to President John F. Kennedy's clarion call to put a man on the moon. Kennedy's goal was for national security reasons: we could not let our cold war enemy, the Soviet Union, win the space race. Was yesterday's national security imperative for a moon shot any greater than today's need to get off Middle East oil? It's hard to imagine that it was.

Today, we import more than 60 percent of our oil, much of it from countries that don't particularly like us. And the import trend is decidedly upward. Inasmuch as we spend roughly half a trillion dollars annually for our national defense, much of which is aimed at fighting the war on terror, we must admit something: we are bankrolling both sides. A hideous, horrible fact is that many of our petrodollars are being deposited in the bank accounts of Mideastern terrorists.

Imported oil is a transportation issue. Most of our imported oil is refined into gasoline to fill the tanks of our SUVs and pickup trucks. Nevertheless, once you start to examine our national energy situation, you realize that our electricity sector is not much better off than transportation. In 2007, we meet our electricity needs mostly with domestic fuels, but that could change if we start to import natural gas in liquefied form (LNG) from the Middle East, as some of our leaders propose.

Aside from potential national security issues, our aged and brittle electricity transmission grid poses a looming homeland security risk. And our dependence on fossil fuels to generate some three-quarters of our electricity creates both financial and environmental risks. After all, fossil fuels are the compressed leavings of dinosaurs and ancient flora. Since they're not making any more dinosaurs, fossil fuels are depletable resources. Moreover, their carbon content makes them major culprits in an environmental crisis of epic proportions.

Yet, most Americans, sad to say, are unaware that we are sleepwalking toward disaster, due both to our electricity choices and the imported oil issue. Imported oil gets attention these days because of our wars in the Middle East. Electricity issues do not receive attention until the lights go out. But they should be addressed before we get to that point, and that's why I'm writing this book.

Introduction

Change is the only constant.

—*Proverb*

Can you believe it was only about 125 years ago that Thomas Alva Edison's invention of the lightbulb was met with public skepticism and fear? In early installations, he hung the following sign near his electric lights:

> This room is equipped with Edison Electric light. Do not attempt to light with match. Simply turn key on wall by the door. The use of electricity for lighting is in no way harmful to health, nor does it affect the soundness of sleep.[1]

That was little more than a century ago—a mere hiccup in time, especially when compared to some of our neighbors in the global community who have thousands of years of history under their belts. And this is just the lightbulb. We can't imagine life without it, yet it has been around for only 100 years or so.

Consider the incredible technological changes of the past 100 years: automobiles, planes, the electric transmission grid, mainframe computers and supercomputers, rockets and trips to the moon, fax machines, cell phones, the Internet, and DVDs, just to name a few. The pace and magnitude of technology innovations in the past 100 years are absolutely staggering.

In the past twenty years, in particular, we have witnessed nothing less than a technology revolution. Who could have anticipated the way that personal computers, wireless technology, and the World Wide Web have transformed our lives? Moreover, who would have anticipated the pace at which these technologies have been thrust upon us, have become part and parcel of our daily lives, and in a very real sense, have affected how we define ourselves?

Some 150 years ago, the Industrial Revolution took off in England. All societies had been powered by biomass (wood, peat, whale oil, and the like) until that time. The Industrial Revolution marked the first time in the earth's history that mankind extracted carbon from the ground (in the form of coal), burned it, and vented the stored carbon into the atmosphere as carbon dioxide.

Today we can harness the power of the sun, the wind, and, yes, even biomass once again. Although fossil fuel energy and nuclear are the only bulk power sources today's Americans have ever known, they're relatively recent actors on the energy stage. And because fossil fuels are depletable, there is mounting moral pressure to reduce our usage and save some resources for our kids and grandkids.

As in the past, the technology scene is changing. Lucky for us, new technologies—powered by vast and renewing resources—have been developed and are ready for marketplace adoption. We're on the cusp of major changes in how we produce and consume electricity. Technologies and distribution systems have been developed. They merely await adoption in the marketplace. But, like the early days of Edison's electric lightbulbs, these technologies and systems face skepticism (or ignorance) from the consuming public and they must overcome barriers in the marketplace. Edison had to deal with the natural gas industry, which, having recently conquered the whale oil interests for dominance in interior illumination, now tried to beat back the electricity interloper.

Change. Like death and taxes, change is one of life's constants. It is not to be feared, although we all do anyway. It's not surprising that those who profit from the status quo, and therefore have a vested interest in it, play on those fears. They capitalize, literally, on our collective fear of change. They suggest that changing our tried-and-true way of generating and transmitting electricity could be dangerous, unreliable, costly. Maybe they're right. Maybe they're wrong. Maybe both.

How is the average consumer supposed to sort through all the rhetoric and make informed decisions? More to the point, does the average consumer care enough to make the effort? Let's face it: nobody but energy geeks cares about the details of energy. Like the child who assumes that milk comes from the grocery store, many Americans think of energy in terms of their direct experience with it: that is, it lights our lights, it powers our air conditioners, it fuels our planes, trains, and vehicles. In fact, much literature on energy refers to the energy of our bodies or our psyches, not to the energy that fuels our national economy and is the lynchpin of our quality of life.

If pressed, the child can tell you that milk actually comes from cows, not the grocery store. Similarly, thanks to headlines emanating from the Middle East, U.S. motorists are aware that gasoline is refined from crude oil and that much of it is imported. Some of it comes from countries with which we are willing to wage war in order to secure our supply of black gold.

On the electricity side, consumers understand that electricity, like gasoline, must be made. While transportation fuels are refined from crude oil, electricity is generated using several fuels: coal, natural gas, processed uranium, falling water, a little geothermal, and, mostly in the Northeast, some oil.

In recent years, utility-scale wind turbines also have contributed to the generating portfolio of some utilities, and the pulp and paper industry generates some of its electricity by using the biomass waste from its own industrial operations. Some larger industries generate part of their electricity from the waste heat of their industrial processes, a technology called cogeneration, which is twice as efficient as power plants.

The systems and technologies we devised in the early and middle years of the twentieth-century served our country superbly. We like to think it's people that make our country great, and that is, without a doubt, true. However, we could not have succeeded on the scale we did without the extraordinary energy resources and delivery systems that fueled our burgeoning twentieth-century economy.

But there is a perverse irony in our twentieth-century success: we've had little motivation to adapt our large, expensive, complex, stressed, and aging electricity system to changing times. Furthermore, we are failing to utilize technologies that became available in the 1990s and at the outset of the current century. In contrast, late-developing countries are using these new technologies to their benefit in the competitive twenty-first-century global economy. They have been able to leapfrog over central station power plant technology and plunge straight into renewable, distributed systems that generate electricity close to the point of use. One consequence is that, unlike us, they don't waste massive amounts of energy generating and transporting electricity to the user.

In terms of our energy situation, the United States is

like a large oceangoing vessel with faulty radar: we haven't detected the icebergs dead ahead and are steaming straight toward them. If we wait until we finally spy them with our own eyes, it will take too long to slow the vessel and turn it away from certain disaster.

What is the nature of this disaster? It's hard to tell with specificity, but no doubt it will have several dimensions. For starters, it will affect our competitiveness in the global economy. Energy is a cost of production, and the United States has the second most energy-intensive economy in the world.* Simply put, we require more energy to produce the same output as other countries that have more-efficient energy systems. This contributes to high costs of production, higher priced products, and loss of market share.

Another dimension of the disaster is that individual consumers will have to adjust their household budgets—not to mention their lifestyles—in order to accommodate rising costs of electricity and fuel. Why do I assume that costs will go up? Because our energy systems are based on shrinking resources—mostly fossil fuels. Once we burn up these resources in our cars and our power plants, they're gone forever.

Some expert observers of the global petroleum scene are sounding the alarm: oil production worldwide is peaking or may already have peaked. It's hard to believe this is possible, just as it was hard to believe M. King Hubbert's prediction back in the 1950s that U.S. oil production would peak sometime around 1970. At the time, this geologist was roundly ridiculed by the U.S. petroleum industry—and who would know better than the industry itself, right?

But Hubbert called it right. U.S. production peaked

* Canada is reputed to win first-place "honors" in this category.

in 1972. Lucky for U.S. consumers, peak U.S. oil didn't impact oil prices. The major U.S. producers moved their operations overseas. We started to drain other countries' subterranean fossil resources just as we had our own.*

What will happen when the entire world's producible oil supplies reach peak and then, by definition, start to diminish? It will be a whole new ball game. Will we run out of oil completely? No. But the immutable law of supply and demand will kick in, and we will pay more and more for our oil. The shock waves will reverberate throughout our economy. Think of all the goods that are part of our daily lives—from plastic bags to cosmetics—and are petroleum-based.

As if the potential economic debacles were not enough, we also face environmental problems resulting from our current energy choices. For example, we know that coal-fired power plants emit thousands of tons of carbon dioxide. This is the key greenhouse gas that traps heat in the earth's atmosphere and contributes in major ways to the unprecedented heating of our planet by dramatically accelerating its natural warming tendencies.

We also know that the fine particulates spewed into the air from these same power plants—as well as from the tailpipes of our cars and trucks—contribute measurably to increased incidences of asthma in both children and adults. Toxic mercury emitted from coal plants falls into our waters, killing fish and poisoning humans who

* Spurred by the U.S. government, some of "the majors" joined with the Saudi Arabian government in the 1940s to create Aramco, the Arabian-American oil company. To persuade the companies to make the move, our government created tax incentives and other financial goodies linked to doing business overseas. Those incentives are in place today, even though one could argue—persuasively—that overseas oil investments are no longer in our national interest.

eat the fish or otherwise encounter the mercury. Another environmental impact is that coal plants require large amounts of cooling water. This is a drain on limited water supplies in arid and semiarid regions of the country.

Yet, coal is the most abundant of our domestic fossil energy resources. To date, it also has been the most affordable in large quantities. Consequently, coal-fired power plants, while only about 33 percent efficient on average, still provide the most dependable and affordable utility-scale electricity of any of our power resources. Even if we wanted to, we could not shut down our coal plants. Nothing exists in today's marketplace that is comparable in both affordability and ability to deliver large amounts of electric "juice."

Fortunately, we have options. As we embark on the adventure that is the twenty-first century, we have an unparalleled opportunity to "have it all" with regard to energy. If we view the first two decades of the twenty-first century as a time of transition in how we produce and use electricity, we can start to phase in the technologies that will become standard and more cost-effective as the years go by: energy efficiency, solar, wind, geothermal, biomass, and, eventually, hydrogen fuel cells.

As these cleaner and more affordable technologies capture more market share, we will be able to back down our use of twentieth-century technologies that require us to burn up resources that can't be replaced. When we use fossil fuels, we use them up. That means they will not be available for those who come after us, for those who might be able to think of better uses for them—for example, in chemicals or pharmaceuticals or materials of some kind. Hogging valuable but depletable resources puts us in the selfish and immoral posture of robbing our children and descendants of their future resources and opportunities.

Sadly, intergenerational equity is an issue that doesn't rise above a whisper in our family-values conversation.

The technology revolution continues, and it's picking up speed. Until recently, renewable energy technologies occupied small niche markets. For example, photovoltaic (i.e., solar electric) cells powered highway warning signs or stock-watering pumps—remote applications that would have required diesel generating units or stringing miles of expensive electric wire. Now, however, there is a growing number of situations in which it is cheaper to use photovoltaics (PVs) than to trench under a street and run a wire to a nearby utility pole.

In addition, improved technologies permit solar electricity to be generated on "thin film" PV materials built into the sides of buildings. This is not science fiction. The next time you're in Times Square, in the heart of New York City, check out the skyscraper at the 4 Times Square address: solar electric is built into the sides of floors 38–45, facing south and east.

Not only are the technologies continuing to be developed and improved at warp speed, but the investment community is starting to take notice and increase its participation in the market. Annual revenues for four renewable energy technologies (PV, wind, fuel cells, and biofuels) grew nearly 39 percent in one year, from 2005 to 2006. One market watcher expects these four technologies to grow to a $226 billion market by 2016.[2] No one is more skeptical of risky ventures or more conservative with their money than investors. When they start to get in the game in a big way, you know these technologies must be legitimate.

Unfortunately, you don't hear much about these power choices. That's about to change.

Part I: The Here and Now

Energy Schmenergy: Why Should You Care?

Energy is an IQ test the American public seems predisposed to fail.

—James R. "Randy" Udall

Energy is the elephant in the room. It underlies almost everything we see and do in our daily lives, and it is a major driver behind some of our thorniest public policy issues. Yet, until recently, we seldom talked about it. As this book is being written, energy finally has become front-page, above-the-fold news, but only because we're at war, primarily for oil. Energy is also starting to receive rightful attention as the key culprit in global climate change.

Not only is energy instrumental in our daily lives, it is key to the future of our country and the planet we will bequeath to our children and grandchildren.

A National Security Concern

Nothing really has taken me aback more as Secretary of State than the way that the politics of energy is … warping … diplomacy around the world.

—Secretary of State Condoleezza Rice[1]

In 2007, the United States meets more than 60 percent of our petroleum needs through imported oil. Roughly a quarter of our imports come from the Middle East, about 15 percent from Saudi Arabia alone. Long considered a U.S. ally in an otherwise unfriendly part of the world, Saudi Arabia sits atop the world's largest oil fields. This humongous resource has allowed our "friend" to serve as the swing producer among the OPEC* members. That is, when we have needed more oil and cheaper prices, Saudi Arabia has been able to open the spigots and produce more oil to meet our needs.

Now several factors are limiting the Saudis' ability to look after our interests in this particular way. Apparently† they are at, or are approaching, their peak oil production capability, thus signifying that they will not be able to fill production gaps in the future. Also, the ruling Saudi family is subject to social and political stresses in their own country and the Middle East. These pressures make their continued friendship with the United States a tricky balancing act. In the coming months and years, this might limit their willingness to continue to act on our behalf.

This last point is critically important and might not be

* Organization of Petroleum Exporting Countries, inspired by and modeled after the Railroad Commission of Texas, the state agency that regulated the rate at which oil was produced from Texas fields back in the heyday of the black-gold rush. As the story goes, a young Venezuelan studying petroleum geology at the University of Texas observed the success of the Railroad Commission in moderating the pace at which oil was extracted, thus conserving the resource and sustaining its market price. Inasmuch as Texas oil production peaked in the early 1970s, these days the Railroad Commission customarily permits all-out production.

† Notice that I use the qualifier *apparently*. That's because OPEC decided in 1982 not to publish production and reserve data. Most of the world's oil resources are owned not by oil companies, but by countries. They believe it's in their interest not to publish data about their oil resources. Consequently, accurate data about the size of the oil resource are very hard to come by, as U.S. petroleum geologists, financial analysts, and Wall Street observers will tell you.

well understood by U.S. consumers: Saudi Arabia is a center of Islamic fundamentalism, and it is a breeding ground for international terrorism. Osama bin Laden comes from Saudi Arabia, as did almost all the terrorists who hijacked U.S. planes on September 11, 2001, and slammed them into the World Trade Center, the Pentagon, and a field in Pennsylvania. In order to remain in power, the Saudi rulers must appease conservative religious interests in their own country. Our petrodollars are part of that appeasement, and thus we, Americans, are bankrolling terrorism. Because of our addiction to oil, we are financing both terrorism and the war on terror.

In addition, the economies of other countries—China and India in particular—are growing, placing them in competition with us for the world's oil. In the past, world supply exceeded demand, providing a safe margin and keeping the law of supply and demand from brutalizing the marketplace. That is no longer the case and never will be again. The law of supply and demand will control world oil prices in ways it never has before.

In the short term, the price of oil may fluctuate, likely within a $15 per barrel range. In addition, OPEC may contrive to set prices low enough to discourage the flow of investment capital to alternative fuels and energy efficient vehicle design. But in the mid- to long term, the price trend will be decidedly upward—just as it has been for domestically produced natural gas in the United States.

Our vulnerability is underscored by the fact that we consume a quarter of the world's oil supply—more than the next five highest consuming nations combined.

Since the U.S. domestic oil supply peaked around 1972 and we started to meet more and more of our energy needs through imports, our national security has become largely synonymous with energy security. Robert

McFarlane, former national security advisor to President Reagan, states that of the $500 billion we spend annually for our defense budget, $80 billion is allocated directly for securing the supply lines for Middle East oil. In addition, the U.S. military spends $100 billion to purchase foreign oil. It's anyone's guess how much of this finds its way to terrorist groups—right out of our own defense budget.[2]

Moreover, McFarlane and other national security experts worry about the fragility of the oil production infrastructure and supply lines coming out of the Middle East. A few mortar rounds placed at any one of half a dozen oil terminals in Saudi Arabia could remove 6 million barrels a day from the market for up to a year. This would cause world oil prices to escalate to more than $100 per barrel, resulting in worldwide recession or depression.

Energy security concerns are related mostly to our transportation fuels. Even the most well-intentioned renewable energy advocates sometimes are confused on this point, touting solar, for example, as our ticket out of the Middle East. Solar and wind are not transportation fuels and do not address our national security dangers.

A Homeland Security Concern

What is the value of energy if you don't have any?

—*John Thornton, "Mr. PV"*

Our country is crisscrossed by networks of pipelines and electrical transmission wires that transport natural gas, oil, and electrons. The main electric transmission system includes more than 200,000 miles of lines and more than 250,000 substations. In the early 1990s, the U.S. Department of Energy (DOE) conducted regional "games" simulating the assorted disasters that could befall our country's energy infrastructure. One such simulation involved an earthquake along the New Madrid Fault that runs north–south through the U.S. heartland. The soft, loamy soil of the broad vestigial Mississippi riverbed would radiate shock waves as far as the East Coast, just as it did in 1811 and 1812 when the fault adjusted itself three times and rang the bell in Boston's Old North Church, some 1,000 miles away. What effect would a similar geologic event have on twentieth-century underground pipelines and aboveground transmission grids?

Similarly, a terrorist act, such as placing sticks of dynamite at the base of a transmission tower on the California-Nevada border, could bring down the entire Western grid. In truth, however, with our aging and brittle electricity infrastructure, nothing as major as an earthquake or as sinister as terrorism is needed to wreak havoc. Because the grid is old and its carrying capacity is maxed out, it doesn't take much to disrupt an entire region's electricity.*

In the past ten years, we have experienced two major failures of regional grids, affecting millions of Americans. The first occurred in 1996 when a squirrel dropped onto an untrimmed tree limb hanging over a wire in Idaho. The system reacted as it was designed to: it shut down in cascading fashion, protecting different parts of the system from the overload created by the event. Satellite photos show the entire western half of the United States in darkness, thanks to a rambunctious rodent.†

The second event occurred in August 2003 when summer heat created sagging wires in Ohio. The rest is history, and mostly a tale of human error. The entire Northeast,

* The national grid is composed mostly of three regional grids that interconnect at a few points. A fourth regional grid covers the far northeastern tip of the United States and eastern Canada.

† Evidently, squirrels have been responsible for any number of blackouts

including eastern Canada, was down for hours on a hot summer day. More than 50 million people were affected and more than 9,000 square miles were without electricity. Among those impacted were Manhattan hospitals that found themselves within minutes of running out of diesel fuel for their backup generators when power was eventually restored.

Whether due to natural disaster, human error, the luck of the draw, or terrorism, our national electricity grid is brittle and vulnerable. It will take roughly a trillion dollars to fully expand and fix it. With better, cheaper alternatives already available or on the horizon, and with our national budget stretched as thin as these wires, an investment in expansion and repair might not get made.

An Operating Cost and a Cost of Living

> *I'm mad as hell and I'm not going to take this any more!*
>
> —*Howard Beale in* Network

I'm always surprised when I hear people say that energy is a fixed cost over which they have no control. Nothing could be further from the truth; energy is a variable cost and it can be managed. To the extent that it is not managed, it is an avoidably high cost.

Energy is embedded in the cost of everything we buy, including food. It is a cost of living reflected not only in our utility and fuel bills. We start each day swallowing energy costs. The price of our cornflakes rises to reflect the

---

over the years. I can't vouch for the veracity of the content, but if even some of the incidents recounted at www.scarysquirrel.org/special/hitension are accurate, hundreds of thousands—if not millions—of electricity customers have been on the receiving end of "nutzy terrorism" at the paws of the "bushytail horde."

increased cost of natural gas–based fertilizer to grow the corn, the higher priced diesel to fuel the farm tractor and delivery trucks, higher energy prices incurred in processing the corn into flakes, and so forth. Inasmuch as food accounts for about 10 percent of the typical American family's household budget, embedded energy costs can add up and cause heartburn.[3]

In manufacturing, the energy input is measured as "energy intensity." It went down 42 percent in the United States from 1973 to 2000 even as our economy grew, meaning that we succeeded in producing more with less power. Nevertheless, due to the continuing inherent wastefulness of the twentieth-century U.S. energy system, our competitors in the global marketplace still use less energy to produce the same output.* Although a number of factors contribute to making our products uncompetitive in the global marketplace, energy intensity is a significant one. It has been reported that we use twice as much energy as Germany and three times as much as Japan to produce the same unit of output.[4]

Take, for instance, industrial motors. They are the single largest electricity consumers in the industrial sector, using 65 percent of all electricity consumed in industry. In many industrial facilities today, there are two speeds—on and off. By replacing them with variable speed motors, industry could reap a 60 percent energy savings (with an impressive return on investment).

Taxpayer-supported institutions (e.g., federal, state, and local governments; school and hospital districts; counties; and so forth) that fail to manage their energy costs create a double whammy for the taxpayers, because

* Because coal-fired power plants are only 33 percent efficient, two-thirds of the energy inputs to power plants are thrown off as waste heat.

tax dollars are squandered to pay unnecessarily high utility costs, and money is diverted from programs and services to pay the bills. A rule of thumb is that at least 20 percent of electricity costs could be saved through attention to the following:

1. Monitoring energy consumption patterns (i.e., continuously looking for opportunities to reduce energy use);
2. Understanding your utility bills (for example, making sure your business or institution is billed at the most favorable rate);
3. Implementing low-cost or no-cost energy saving measures; and
4. Behaving in energy-conserving fashion.

A Business Risk

It is not the cost of electricity that drives our decision-making process; rather, it is the cost of NOT having electricity.

—*Jeff Byron, energy director, Oracle Corporation*

Twentieth-century electric utilities prided themselves on usually measuring up to their self-imposed standard of 99 percent reliability. This meant that, although the power might go out occasionally, the electrons would flow 99 percent of the time. They set the bar astoundingly high, one would think. In the twenty-first century, with our increasingly electricity-dependent economy, believe it or not, that standard is not high enough.

IT-dependent businesses require "six 9s" of reliability from their electric utilities. That's because of the enormous costs of lost business due to power glitches. Think of your own experience: what do you do if, for example, you log on to Amazon.com and their system is down? You head for

Barnesandnoble.com (or vice versa). Stock brokerages suffer losses of $5 to $7 million every hour if their electricity-driven systems go down. Credit card services, $2 to $3 million. Businesses nationwide lose $35 to $70 billion every year due to power outages.

Yet, we continue to gamble, patching up and expanding our massive brittle transmission system, even when we hear that a mere squirrel can bring down the entire Western grid. We invest in fire insurance every year, even though fewer than 1 percent of us—thankfully—will lose our homes to fire. Similarly, we could decide to invest in on-site electricity generation and backup as insurance against increasingly likely failures of the national grid and regional distribution systems.

The price of energy is a line item in our family budgets, but it is a genuine risk for business. Energy customarily amounts to no more than 5 or 10 percent of operating costs in non-energy-related businesses. Nevertheless, during the natural gas crisis of 2000, small firms found that the increased cost of energy equaled or exceeded their slim profit margins. Similarly, rural hospitals have known for years that the cost of energy can be the difference between keeping the doors open or shutting them forever and sending their rural patients to city hospitals, often located some distance away.

Our mothers warned us not to put all our eggs in one basket. Our financial advisors, as well, tell us to diversify our stock portfolios. Why wouldn't you and your utility, in similar fashion, diversify your electricity portfolio?

Right now, we don't have a lot of choices when it comes to transportation fuels. We do, however, have choices in electricity resources. In addition to the twentieth-century fuels such as coal, natural gas, and processed uranium, we have utility-scale wind, concentrating solar power (CSP), and the least expensive, most abundant, untapped domestic clean energy source of all: energy efficiency.

Yet, in any single utility service area, you are likely to find that one fuel or another dominates your utility's generating portfolio. This puts you at risk, because, if the price of that fuel goes up, your electricity prices will rise accordingly and you will be held captive to rate increases. For many Americans, it might mean a cancelled family vacation. For people on low or fixed incomes, however, this could be disastrous—even deadly, as when elderly people turn off their air conditioners or heaters to save money. For a small business, it could mean bankruptcy.

Individual consumers can make energy choices that affect their electricity consumption and costs. They can turn off lights and unplug loads such as computers and TVs when they leave home; they can purchase compact fluorescent lights (CFLs); they can invest in renewable energy, if it makes sense in their individual situation; and they can participate in their utility's voluntary green-power purchasing program. In these and many other ways, individuals can hedge against the risk of rising electricity costs.

A Force in Local Economies

> *Converting the wind into a much-needed commodity while promoting good jobs, the Colorado Green Wind Farm is a boost to our local economy and tax base.*
>
> *—John Stulp, commissioner, Prowers County, Colorado*

In the 1980s, Nebraska's state energy office published a document encouraging citizens to offset the need for energy by using it efficiently. What distinguished this message, however, was that it exhorted Nebraskans not to send their dollars to the "blue-eyed Arabs in Texas," but rather to keep their money at home in their local economies.

Aside from this unfriendly characterization of fellow Americans, the Nebraska publication heralded a new kind of analysis: examining and quantifying the effect of our energy choices upon local economies. Several years later, the town of Osage, Iowa, conducted its own analysis.

Osage learned that each dollar spent locally on general merchandise (e.g., at the local grocery or clothing store) generated about $1.90 in local economic activity—almost double the amount of the expenditure. That's because the dollar spent locally created local jobs, both directly and indirectly, and the people thus employed spent their money locally, multiplying its impact throughout the local economy.

Osage found that $1.00 spent on petroleum products generated $1.51 of local economic activity (for example, the local gasoline franchise holders and their employees). Utility services generated $1.66. Energy efficiency created a multiplier of $2.23. That's because energy efficient products—insulation, caulking, high-efficiency appliances, and so forth—required local labor to sell, distribute, and install. Energy efficiency products and services are labor intensive (as are distributed renewable energy products such as solar water heating, but they weren't included in the Osage study).

In addition to developing an industry beneficial to the local economy, Osage created a selling feature for the town itself. Town leaders persuaded desirable businesses to relocate to their town, attracted by the reduced operating costs resulting from Osage's investment in energy efficiency.

A Key Culprit in Global Climate Change

Never mind what you've heard about global warming as a slow-motion emergency that would take decades to play out. Suddenly and unexpectedly, the crisis is upon us.

—*Jeffrey Kluger*[5]

Among the greenhouse gases, carbon dioxide is the chief contributor to global climate change.* In the United States, electricity generation accounts for 40 percent of carbon dioxide emissions, making electricity generation the largest source of global climate change.[6] The other key culprits are nitrous oxides, oxides of sulfur, and methane. These emissions are related in large part to our agricultural activities; however, they also are related to energy production.

All are emitted into our fragile atmosphere in the production and consumption of fossil fuels. If you look at historical data of atmospheric concentrations of greenhouse gases, concentrations rise dramatically and abruptly, starting around the time of the Industrial Revolution. Picture a graph on which the vertical y-axis measures concentrations of greenhouse gases and the horizontal x-axis marks years. Picture the J curve of data points jumping abruptly upward, roughly coincident with the Industrial Revolution of the 1800s. This is the famous "hockey stick" graph. It is especially dramatic when plotted for carbon dioxide.

It seems breathtakingly ostrichlike to me, but in the year 2007, there still are people who concede that the planet is warming but question whether or not the causes

* Often the term *global warming* is used. However, the impacts of the buildup of greenhouse gases vary in different regions of the world, and some places actually may experience cooling. For sure, one predicted result is extreme weather patterns—which could mean more snow in some places. Consequently, I prefer to use the term *climate change.*

are largely anthropogenic. My own opinion is, we did it.

We all know that our home planet has natural warming tendencies. However, the hockey stick graph illustrates in dramatic fashion how, starting with the Industrial Revolution, we transferred carbon (embedded in coal) from under the ground, oxidized it by burning it, and catapulted it in gaseous form into the atmosphere. We're still doing it. Currently, electric utilities across the country have more than 100 coal plants on the planning books.

The fact of the matter is that earth's atmosphere is extraordinarily thin and fragile. If earth were the size of an apple, our atmosphere would be the equivalent of the protective peel. Gone forever are the days when mankind was so insignificant and the planet so large that we could do anything to our immediate locale without fear of lasting consequences. One of the first voices to sound the alarm was Rachel Carson. She said, "Only within the moment of time represented by the present century has one species—man—acquired significant power to alter the nature of his world."

Making matters worse, by orders of magnitude, are feedback loops, a phenomenon unheard of in the popular press until recently. Polar glaciers are melting at a rate far faster than predicted. That's because of one of the feedback loops. Snow and ice reflect sunlight back to the atmosphere, deflecting a good part of the heat of our solar furnace. This is the albedo effect. As glaciers start to melt, however, the darker colors of land and water absorb more solar heat than the snow and ice would have. The areas of exposed land and water grow, further accelerating the pace of melt. Astoundingly, at least one glacial ice sheet is now retreating (melting) at the rate of five feet every hour. Tragically, this lends whole new meaning to the expression "at a glacial pace."

In addition, no one knows exactly how all the freshwater glacial melt will impact earth's saltwater oceans. Earth actually has a kind of circulatory system, called the thermohaline cycle. It's a phenomenon that creates strangely warm currents off otherwise cool land masses and keeps places such as England and Ireland from becoming uninhabitable during the winter.

Similarly, no one knows what the impact of the assorted consequences of global climate change will be on the inhabitants of the world's oceans. Rising water temperatures, increasing acidity, increased evaporation rates—the list is long and terrifying. The many systems are complex and interrelated. Predictive models, while improving, do not begin to capture the many mechanisms and feedback loops. This is why some of the early climate models in the 1960s and 1970s predicted global cooling rather than warming. Today's climate change naysayers use the errors of those primitive early models to justify their skepticism.

A Regional-Environment and Human Health Issue

> *Tug on anything at all in the universe and you'll find it connected to everything else in the universe.*
>
> —*John Muir*

Aside from their impact on climate change, power plants account for two-thirds of our SO_x emissions and one-third of NO_x emissions (implicated in smog). The tall stacks of coal plants shoot these emissions into the sky and sometimes into the jet stream. The emissions can mix with moisture to create a toxic potion of acid rain that, if transported by the jet stream, can fall far from the point of origin. This is what happened with the fish kill in Lake George.

The nation's 600 coal plants also contribute more than 40 percent of all mercury emissions in the United States, discharging 48 tons into the air every year.[7] If mercury falls into water, it's ingested by fish. Then we eat the fish, and mercury is in the food chain. Mercury pollution of industrial origins was identified as the cause of a 1955 disaster in Japan, where women who had eaten mercury-tainted fish gave birth to babies with mental retardation and other abnormalities. Mercury has since been linked to nervous system damage in fetuses and children. In the United States, all but five states have issued health advisories warning of potential mercury contamination in fish.

In its vaporous form, mercury is absorbed through the lungs. Similarly, humans inhale the fine particulates thrown off by coal plants. These are known to constitute a major health threat. Asthmatics and the elderly are particularly susceptible. Long-term exposure is thought to cause cancer. Exposure to these airborne toxins is highly dependent on weather patterns, with higher risk to human health occurring in warmer weather.

The combustion of fossil fuels—whether to generate electricity or to move vehicles—produces emissions and toxins that contribute to or cause any number of ailments. The American Lung Institute has attributed more than $50 billion of health care costs every year to the health impacts of our energy choices, noting that almost half of the population lives in areas that are not in compliance with national ambient air quality standards.

An Underlying Driver on Wall Street

Stocks decline as oil prices rise.

—The Denver Post *headline for February 24, 2007*

When the price of oil goes up, it's an inflationary factor, because oil ripples through our entire economy. Don't think just of fuel. Think of all the things that are made from petroleum—chemicals, garbage bags, plastics, cosmetics, fabrics, pharmaceuticals. Oil is everywhere, though we never think about it. Consequently, when the price of oil goes up, the price of everything goes up. The result is that disposable income goes down when the price of oil goes up.

That's why the issue of peak oil is so important and it's why, in writing this, I can almost cut and paste from the national security discussion to this section. Although we lack accurate data about the precise size of the world's oil resource, petroleum engineers and Wall Street analysts who monitor key indicators conclude that we are at, or very near, global peak oil production.

Does that mean that we will run out of oil the day after the peak is reached? No. But it will cost a lot more.

Picture a graph, with oil production measured on the vertical y-axis and years ticked off on the horizontal x-axis. In general, oil production has been thought to look like a classic bell curve, with production ramping up over time, peaking at some point, and then declining at roughly the same rate at which it ramped up.* The peak might look like a mesa or it might be a sharp point, depending on the characteristics of the individual oil field.

* In fact, the rate of decline in some fields might not be as moderate as a bell curve. Fields that have been overproduced—that is, produced at a faster rate than geologically prudent—may experience a more abrupt decline. Graphically, it might look more like a Thelma-and-Louise curve than a bell. This is frightening, because we suspect that our historic friends the Saudis have overproduced some of their fields in order to meet our voracious needs. The decline of those desert mega fields might be more precipitous than anticipated.

This could mean that we might be at peak for some time and not know it. Matthew R. Simmons, the renowned Wall Street analyst who is among those raising the alarm about peak oil, compares our global thirst for oil to that of your family car: You can be cruising along at 70 miles per hour until the moment when your car runs out of gas and limps to a complete halt.

The only thing that saves you from running out of gas in the middle of nowhere or—worse yet—rush-hour traffic, is the gas gauge. Inasmuch as we lack accurate data about the world's true oil resource, the only way we will know that we're at peak is through upward pressure on prices. World demand for energy is projected to grow at a rate of 2.2 percent annually, up from 1.6 percent during the past decade.[8] As demand grows and supplies do not, the law of supply and demand will kick in and prices will rise.

A number of experts think we are already there. Wall Street analyst James J. Cramer says, "[There is] an inability to find any new oil of consequence coupled with a voracious worldwide demand … To me, this $60 [per barrel] level is becoming the norm. … Get used to it."[9]

Legendary Texas oilman T. Boone Pickens concurs. He says, "There's still oil to be found, but not in the quantities we've seen in the past. The big fields have been found, and the smaller fields, well, there's not enough of them to replenish the base." Pickens predicted $70 per barrel in 2007.[10]

The reason I'm talking about this—even though America's addiction to oil is mostly a transportation issue—is because of the devastating impact that dramatically increased oil prices are likely to have on our national economy. And here's a scary thought: other than changing our energy-consuming habits, there isn't much we can do to control this inexorable march toward economic disaster.

Here's the good news: we *can* change our energy ways. The technologies and resources exist for us to use energy efficiently and to start the switch to clean, domestic, renewable energy resources. These alternative technologies are affordable. Moreover, if we accounted for all of the costs of today's energy technologies (e.g., military costs to secure supply lines, environmental and health costs associated with coal, and so forth), the alternatives would actually be cheaper than the conventional choices. The bottom line is that we need to start thinking—immediately and seriously—of the cost of *not* changing our energy ways.

Electricity 101: Power, Yesterday and Today

> *The utility industry's efficiency has not increased since the late 1950s. … Americans pay roughly $100 billion too much each year for heat and power. Put another way, the typical utility consumes three lumps of coal to deliver one lump of electricity.*
>
> —*Richard Munson*

One absolutely can not overstate the incredible scientific and engineering marvel that constitutes the United States' electricity generation, transmission, and distribution system. It was created in 1882 by Thomas Edison and employees of his Edison Electric Illuminating Company of New York. One of his employees, Samuel Insull, struck off on his own and created the forerunner of today's transmission and distribution system. While Edison believed in competition, Insull was happier being insulated from that kind of business pressure and was able to sell politicians on the idea of regulating electric utilities as natural monopolies.

Incredibly, considering all the other technological revolutions of the twentieth century, the incandescent lightbulb and the electricity generation and transmission system remain much the same as when they were created in the nineteenth century. Incandescent lightbulbs are still only 10 percent efficient in creating light, throwing off 90 percent of the electricity input as waste heat. Early generation and distribution systems were only about 10 percent efficient as well. In the mid-twentieth century, utilities succeeded in improving generation efficiencies to 33 percent, where on the whole they remain to this day. Despite the incredible advances in telecommunications and other technologies that might be of use in our electricity system, utilities have not advanced the system in significant ways. Perhaps this is because their profit margins are protected by their regulators, and because they lack market incentives to invest in technological improvements. Perhaps regulators have not nudged them to invest in infrastructure or R & D. Perhaps all of the above.

In any case, whether you pay your electricity bill to an investor-owned utility (IOU is the acronym, fittingly), to a municipally-owned utility, or to a rural electric cooperative, you probably think you're paying too much. Maybe you are, and maybe you aren't. The bottom line, though, is that no matter who charges you for electricity, they all conform to one general model. They all generate electricity—whether from coal, natural gas, nuclear fission, hydropower, or oil—at a central station power plant. The voltage is stepped up for greater efficiency in transporting the electrons over long distances and is transmitted through high-voltage wires to transmission substations, where the voltage is stepped down. From there it goes through more wires to distribution substations, where it is stepped down again before it is sent via feeder lines into our neighborhoods, homes, and businesses. The final step-down occurs in your immediate neighborhood at a transformer. If electricity came into your home or business at too high a voltage, it would fry all of your electricity-using stuff—from appliances to computers—in a heartbeat.

Without caring much about any of this, you're able to flip a switch or punch a button and turn on the myriad devices that help simplify or enhance your life. And you do it for, speaking in glittering generalities, between roughly 6 and 19 cents per kilowatt-hour (kWh). It's the bargain of the twentieth century.

Your utility has a very hard job to do, and U.S. utilities have been doing it extremely well for a long time. They have to anticipate how much electricity you're going to use, and then be ready to supply it when you need it. This is called capacity. If you live in a state that has not deregulated its utility sector and that is subject to regulatory oversight through your public utilities (or public service) commission, your utility is required to build adequate capacity and supply enough electricity to everyone who lives in the utility's service territory. This is called the obligation to serve.

The need for capacity is why your utility builds power plants or contracts with electricity providers. It takes a lot of energy to produce this capacity and it takes a lot of energy to deliver the electricity from generation to point of use. Some 25 percent of our fossil fuel consumption in the United States is related to the production and delivery of energy or electricity to where we use it. Ten percent of the electricity generated at power plants and shipped over the wires is lost due to resistance over distance. During peak congestion times on the wires, the line losses double. All that electricity, gone. Just gone.

Utility planners are like the chef at your local eatery. Let's call it the Demand Café. Ideally, the chef wants to purchase exactly as much food stock as you will select from the menu. She doesn't want to waste money by buying stock that will spoil for lack of consumption. At the same time, she doesn't want to run out of the ingredients for your favorite dish, thereby perhaps driving you to the Request Café down the street.

Picture a graph, with the vertical y-axis measuring the amount of electricity generated and the horizontal x-axis ticking off the hours in a twenty-four-hour cycle. Base load is the amount of electricity your utility knows for certain you will need (demand) and it can sell. It occurs throughout the day and night, producing a shape on our graph like a prone rectangle.

However, there are times within a twenty-four-hour cycle when not as much electricity is needed—in the middle of the night, for example. Utilities call that a "valley" in their load shape. They hate valleys, because they're paying to generate electricity that isn't being bought—like the chef who stocks up on too much lima beans and liver.

Similarly, there are times when your demand for electricity exceeds the base load amount the utility had planned to sell you. On hot summer days, for example, a predictable peak may start in the late afternoon, when the day's heat is accumulating and air conditioners ramp up. It continues into the early evening, when folks get home from work, turn on their air conditioners, fire up the ovens to cook dinner, turn on the TV and stereo, and flick on those awful incandescent lights.

Like the chef's dinner rush, utilities anticipate these peaks. However, it's not economical for them to build massive base load power plants just to meet these spikes in demand, knowing as they do that the power plants will continue to spew out unneeded—and unpurchased—electricity for the rest of the twenty-four-hour cycle. Consequently, utilities either purchase this peak power from someone else or they build smaller power plants, "peakers," that they can ramp up to generate electricity quickly—within a matter of minutes. This has technology implications: coal plants can't be ramped up quickly; natural gas plants can.

When you think about it, the ideal load shape for a utility is a flat line, running horizontally over that imaginary graph and extending across a twenty-four-hour period. Like the chef, the utility would like to know exactly how many people will drop in at the café and what they will order. This would be the planner's dream. It will never happen.

This is today's—and yesterday's—model. As superbly as it served us in the twentieth century, a superior model is available to us today and tomorrow, due to technological advances. In order to know where we're going, however, it's instructive to understand where we've been. All of the examples that follow produce bulk power at central station power plants. That is, electricity is generated in huge amounts at a central point. Then it's delivered via wires to your home or business. The delivery system is the same for all generating fuels. What varies are the fuel and power plant technology. I will not deal with all twentieth-century technologies, just the big three that supply about 90 percent of our electricity: coal, nuclear fission, and natural gas. Oil, hydropower, and some biomass make up most of the rest, with wind starting to take some utility market share.

Coal

They don't call it King Coal for nothing. Coal is the most abundant fossil fuel in the United States, and we sit on 27 percent of the world's proven reserves. Although we used up a lot of it in the past century, a lot more remains. The law of supply and demand being what it is, the sheer abundance of coal makes it seemingly affordable. Coal has been, and continues to be, the workhorse of our electricity-generation system. Across the country, about half of the United States' electricity is generated by coal. In many utility service areas, however, the proportion of electricity generated by coal is much higher.

Coal plants are massive, requiring massive infrastructure investments, massive rail networks, and massive coal supplies. Thus, they leave a massive footprint. To give you some perspective, consider one utility in Wyoming. This utility supplies 1.6 million homes across its territory with electricity. To do so, it fires up three generating units and burns 24,000 tons of coal every day.[1] That's *one* utility.

Unless a power plant is located at the mine mouth, the coal must be transported to the power plant, where it's burned. Most coal is transported by rail, in big open bins, hooked to maybe 99 or 149 other open bins piled high with their dusty black cargo. Several locomotives spew diesel fumes as they pull the indescribably heavy load. Anyone who has ever waited at a railroad crossing for a coal train to pass knows how long those trains are. When they reach their destination, they dump the coal, which sits on-site in huge open piles until it is conveyed into the plant and burned up. Depending on the size of the power plant, one or two or more coal trains might be required every week to deliver sufficient fuel.

It takes from one or two days to as long as a week to ramp up a coal plant and recalibrate it with the system if, for some reason, it goes down. The mounds of coal that sit outside are fed into the plant and enable it to generate a steady, sustained amount of electricity at all times—whether it's needed or not. Once you "turn on" a coal plant, you can't adjust it like you can a thermostat, and you don't want to turn it off unless you absolutely have to, because it takes a long time to bring it back on line.

The Good

Coal is used to generate our base load electricity. That is, it provides most of the electricity that utilities anticipate they will need for most of every day, day in and day out.

First on the list of positives is the fact that coal is relatively affordable. Electricity generated from older coal plants costs around 2 to 3 cents per kWh. Other than hydropower and energy efficiency, this is the cheapest electricity resource we've got, in large part because these older plants are fully depreciated.

Second, coal is seemingly abundant. The coal industry avers that, at the current rate of use, the 270 billion tons that have not yet been mined will provide the United States with something approaching a 250-year supply.

And finally, we don't have to go to war over coal. It's a domestic resource.

The Bad

Have you ever held a lump of coal? It's dirty. And the dirt that rubs off on your hands is the least of it. You see how dusty it is. When coal is burned, fine particulates are emitted that can clog your lungs when you inhale them. This can cause severe respiratory ailments and has been linked to increased incidences of childhood asthma in communities downwind of coal plants.

Moreover, coal plant emissions contain toxins. There are several pathways for mercury to impact human health. No matter how the exposure occurs, mercury has been linked to afflictions of the central nervous system, especially in children and fetuses.

Aside from health impacts, coal plants affect the communities in which they are located in other ways. An issue that is growing in importance is that coal plants consume a lot of water for cooling purposes. New technologies have reduced this amount somewhat, but it's still significant. According to DOE's National Energy Technology Laboratory, each kWh of electricity requires about twenty-five gallons of water to produce. In this regard, we use roughly as much water turning on lights and running appliances as we do to take showers and water our lawns.

Coal-generated electricity requires massive amounts of land area for the power plant and for storage and handling of the mounds of coal that are fed into the plant every day. Hundreds of coal cars must come on-site weekly to disgorge their black loads. This requires railroad tracks and multiple sidings. As the 100- or 150-car trains grind slowly and loudly through town on their way to and from the power plant, traffic sits and waits—vehicle engines idling and emitting exhaust fumes, and motorists sitting unproductive (and perhaps fuming) in their vehicles.

Most important of all, coal plants emit greenhouse gases—particularly carbon dioxide, the leading climate change culprit. Coal plants have been identified as a key factor in global climate change, accounting for an estimated 40 percent of the United States' carbon dioxide emissions. Twenty-first-century "clean coal" technologies to reduce greenhouse gas emissions are under development. However, they're not yet proven, affordable, or commercially available.

These varied impacts of coal use are captured in an umbrella term, *externalities.* Although you don't pay for the energy-related health care costs in your utility bill, you pay in other ways. You pay in increased insurance premiums, in increased health care costs, in taxes to pay the costs of environmental monitoring. You don't directly pay the price of folks sitting in their cars and waiting for coal trains to rumble through. But, in the aggregate, society pays. We pay in terms of the lost productivity of those drivers, and we pay in terms of the wasted gasoline and the tailpipe emissions from their idling engines.

Any discussion of the relative costs of different electricity generation fuels rightly should include estimates of these externalities, but it doesn't. That's because it's extraordinarily difficult—not to mention contentious—to assign proxy values to the impacts of energy production and use. Although we can't agree on values for the different impacts, all parties can agree on one thing: zero is the wrong value. Yet, because we fail to deal with the externalities issue, that is the only value that has been assigned to date.

In addition to adverse environmental and health consequences, future affordability is also questionable. First, consider how we get coal. Mostly, we ship it via rail from mine mouth to power plant. This creates a significant and growing cost factor. Our railroad tracks are old, in disrepair, and packed with so much coal dust in some places that it derails entire trains. Today's tracks are insufficient to carry the thousands of coal cars that must run from mine to power plant and back every day.[2]

What happens when there's a derailment or other event that prevents the coal from getting through? You still need electricity, and your utility still has the obligation to get it to you. So the utility fires up its natural gas peakers or purchases expensive coal or natural gas on the spot market. When this has happened in recent years, consumers paid tens of millions of dollars in increased utility bills—raising individual bills by 10 percent or more.

It gets worse. As of 2007, utilities across the country are planning to build more than 100 additional coal-fired power plants, both to meet our growing electricity needs and to substitute for natural gas–generated electricity, which has become too expensive. To upgrade and expand the rail infrastructure will require significant investment. Those who must use the rail system are captive customers of the six rail companies that survived the rail industry consolidation of the 1980s. They—we—will pay for this infrastructure upgrade. We also will pay more than $100 billion for the planned new coal plants.

Because of these uncertainties, some electric utilities are not sure they can secure sufficient amounts of coal to fuel the power plants that are on the planning books. Unbelievably, in some cases utilities even admit uncertainty about their ability to fuel *existing* plants.

One thing is certain, though: utilities enter into long-term, fixed-price contracts for coal. When those contracts expire and come up for renegotiation, the sticker shock for consumers will be dramatic. Today's spot market prices are a good indicator of how high future contract prices will be, and in recent years the spot market prices have doubled and even tripled.

Abundance has been key to coal's affordability. But how abundant is it really? Years ago, Dr. Al Bartlett, a professor of mathematics at the University of Colorado, calculated that—due to the exponential function and our growth in both population and energy-consuming behaviors—the United States' energy demand doubles about every seven years. Consequently, there is no such thing as "at present rates of consumption," which is the qualifier used by those who assert that we have enough coal to last 250 years.

To illustrate the power of the exponential function, Bartlett used the following riddle: Imagine you are a bean in a big jar. Start with just one bean at 11 A.M. and double the number of beans every minute until the jar is full one hour later, at noon. What time is it when the jar is half full?

If you guessed 11:30, you're in good company—almost everyone says that. But it's wrong. The correct answer is 11:59. Remember: the number of beans doubles every minute. You know the jar is full at noon. By definition, going backward, the jar is half full at 11:59.

The reason I asked you to pretend you're a bean is to put yourself in that environment. At 11:59, everything looks fine to you. The jar is only half full. You have lots of bean friends, your environment is spacious, and you have plenty of room to grow, because the jar is half empty. One short minute later, BLAM! The jar is full, you're squeezed by all the other beans, and it's too late to do anything to fix the situation.

This illustrates the power of the exponential function. As Bartlett says, "The single greatest shortcoming of the human race is the failure to understand the exponential function."

Apply the exponential function to our coal supply. Factor in the added coal plants that are on utility planning books in 2008, plus the growing interest in liquefying coal for transportation fuel. You have run headlong into the exponential function. You also have hit the law of supply and demand. You can expect that coal will be neither abundant nor affordable as the twenty-first century progresses. Another factor to consider is that U.S. production of high-quality coal—that is, coal having high energy content—peaked around 1990.[3]

Moreover, the cost of materials needed to construct power plants of any kind—steel, cement, and so forth—is rising dramatically. One would expect this as the normal cost of living increases. The situation is exacerbated, however, due to the globalization of our twenty-first-century economy. Steel that otherwise would have been ours has been exported to the growing economies of China and India, dramatically increasing the price.

Coal is a depletable, nonrenewable resource. Any that we take today will not be available for our children and grandchildren. We expect them to be smarter than we are, and they're likely to find higher uses for it rather than merely burning it up to make steam to turn generators. When evaluating our energy choices, intergenerational morality needs to be included in the discussion.

The Balance

Coal-fired power plants are capable of generating the massive amounts of electricity we need to power our increasingly electricity-dependent society. Even in the face of rising costs, coal likely will be significantly cheaper than nuclear or natural gas.* Utility-scale renewable energy technologies are promising, but, other than wind, they're not widely available or cost competitive today.

Coal will continue to provide a massive and essential amount of base load electricity well into the twenty-first century. There are no immediately available, affordable, and reliable alternatives that can generate electricity at this scale.

Nuclear Fission

Stupendous amounts of energy are emitted when atoms are split in a process called fission.† We make weapons of mass destruction with this technology. We also use the heat it produces to heat water to produce steam and run electricity generators. Today, 103 commercial nuclear power plants in 31 states generate about 20 percent of the

* Unless it becomes subject to a carbon tax or other financial penalty intended to mitigate its climate change impacts.

† Note the distinction between fission and fusion. Today's technology is fission, which entails splitting atoms. Fusion is when atoms are fused together, duplicating the technology of the sun, which, to date, is the only nuclear fusion generator in our earthly experience. World-class scientists have been working for at least fifty years to try to harness that kind of energy. Fusion always seems to be "fifty years out." Has been for the past fifty years, still is today.

United States' electricity. Some states rely heavily on this base load generation technology, with New Jersey coming in first at 52 percent. Nuclear power supplies 78 percent of France's electricity.

No nuclear power plants have been licensed in the United States since 1978. However, government incentives, technology improvements, "stranded assets,"* and concerns about global climate change are prompting the consideration of thirty new reactors.

The Good

Nuclear fission can generate the massive amounts of base load electricity we demand. Moreover, if we build additional nuclear power plants, and if sufficient supplies of the uranium fuel are available and affordable, this technology can supply any reasonably anticipated future needs, and it can do so reliably.

Unlike fossil fuels, which are burdened with the built-in carbon that is the legacy of their origin, nuclear is clean from a greenhouse gas perspective. As public understanding of the dire consequences of global climate change grows, it is probable that some kind of financial penalty or restriction will be attached to fuels that emit greenhouse gases. Should this happen, the price of nuclear, by comparison to the penalized fossil fuels, might be cost competitive or at least not prohibitively more expensive.

* Stranded assets are equipment or facilities that continue to exist on utilities' books, even if they're not in use or earning a return. They represent sunk investments of one kind or another. In the process of deregulation, utilities were compensated for some of these assets, even though they might have been mothballed or merely on the planning books. Some utilities, Tennessee Valley Authority among them, are bringing previously abandoned nukes (nuclear power plants) back on line.

The Bad

"Too cheap to meter." In the 1960s and '70s, proponents predicted this would be the cost of nuclear. They were wrong. When U.S. utilities invested in nuclear as a means of getting off oil after the 1973 Arab oil embargo, it appears that, technologically, they didn't fully appreciate what they were getting into.

I don't think it's unfair to say that neither utility executives nor their government regulators possessed an adequate understanding of this complex technology and the difficulty of weaving it into utility-management systems and corporate cultures. In addition, vendors—competing for market share—promoted utility-scale nuclear reactors without waiting for the operating experience that would have informed both system design and operational handling. Simply put, utility-scale nuclear was not ready for prime time in the 1960s and '70s, when utilities started to substitute it for oil.

Nuclear fission is an extremely complicated way to heat water to produce steam and power generators. A large nuclear plant has some 40,000 valves—ten times more than coal or oil plants of similar size. In addition, fuel rod temperatures reach a searing 4,800 degrees,[4] placing significant demands on materials science to withstand this brutal thermal punishment.

In hindsight, it seems there was a willingness to trust technology, almost blindly. I remember being escorted by utility executives on a VIP tour of a nuclear power plant under construction in the late 1970s. When we came to the mock-up of the control room, several of us asked about the employees who would monitor the many readouts and displays on the control panel. Would they be PhD physicists, trained to understand the equations and theories behind the indicator lights? With almost a literal pat to the top of our heads, the utility executives explained that there would be no need for employees to understand the science behind the panel. In fact, while they would be specially certified for their jobs, they probably wouldn't need much more than a high school education; the technology would run itself.

About six months later, in 1979, a valve stuck open in one of the reactors at the Three Mile Island Nuclear Generating Station near Harrisburg, Pennsylvania. Primary coolant was lost, causing a partial meltdown of the core. This is known as the China Syndrome, so named because a nuclear core meltdown theoretically could generate enough heat to melt through the earth to countries on the opposite side of the globe. It took years to sort through what had gone wrong at Three Mile Island, and it took fourteen years and $1 billion to clean up the mess at the problematic Unit 2 reactor.[5]

In the 1980s, there was a far worse disaster at Chernobyl in Russia's Ukraine region. Leaked radioactivity was blamed for hundreds if not thousands of deaths, as well as the births of mutant animals such as three-legged calves. Regardless of the horrifying hype, the region is still feeling the effects of this terrible accident.

The impact in the United States, however, was that regulators at the Nuclear Regulatory Commission started to require redundant safety systems and other features aimed at preventing another Three Mile Island or Chernobyl. Huge cost overruns and lengthy construction delays became the norm, resulting in the bankruptcy of one utility in the Northwest. The utility's unfortunate acronym was WPPSS (Washington Public Power Supply System). Pronunciation of it sounded like "whoops," and it became the synonym for the United States' nuclear industry.

Because of the significantly increased construction

costs, nuclear power has been anything but too cheap to meter, coming in between 2 cents per kWh (at old, depreciated plants with a high capacity factor) to as high as 15 cents per kWh.[6]

Moreover, this is just the cost of the electricity itself. The costs of externalities are not included in these figures. It doesn't count the cost of liability insurance, or the costs of transportation to waste disposal facilities, or the cost of waste storage. These are huge.

Lest you think you don't pay those costs, let me assure you that you do. You pay, in part, through your electricity rates. Importantly, however, you pay for these externalities through other venues, primarily taxes. Because fissionable materials can be stolen from power plants and weaponized, you pay, somehow, for any associated risk in our post-9/11 world.

You also pay for the inherent risk associated with this technology. Although the number of "events" at nuclear facilities has been relatively small in their thirty years of operation, and although utilities have a fairly solid safety record with this technology, the potential consequences of a nuclear accident are grave and overwhelming. To the financial community and insurance actuaries, the financial risk assigned to nuclear power plants is great: critical events are relatively unlikely, but if they should occur, the consequences would be dire.

Liability issues would have nipped this technology in the bud, but for timely government intervention in the marketplace. Around the 1950s, Brookhaven National Laboratory monetized some of the risks and concluded that a nuclear disaster could cause 4,300 deaths, 43,000 injuries, and $7 billion in property damage.[7] Utilities could not have afforded the liability and would not have embraced the technology.

Consequently, in 1957, Congressman Melvin Price and Senator Clinton P. Anderson authored legislation that absolved utilities of significant liability in the event of a nuclear accident. They created a $560-million fund, capitalized with taxpayer dollars, to cover any damages regardless of a utility's carelessness or culpability. Under the Price-Anderson Act, the American taxpayer continues to pay this insurance on behalf of our economically "mature" electric utility industry.

An assessment of nuclear power yields a very expensive bottom line.

We would be remiss if we didn't consider the fact that uranium, which is processed to produce nuclear fuel, is a depleting resource. It was mined heavily during the Cold War. Now the resource is under further stress, as China, India, and Russia plan new nuclear power plants. The cost of uranium is trending sharply upward, from ten dollars per pound a few years ago to eighty-five dollars a pound in 2007.

In America, just as no new nuclear power plants were permitted after 1978, there was also no investment in new uranium mines or in processing facilities. Worldwide, uranium production is adequate to meet only 65 percent of current reactor requirements, much less planned growth. Currently, we rely on Russia for half our nuclear fuel, under a "swords to plow shares" program that converts some 20,000 Russian nuclear weapons to fuel. That agreement will end in 2013.[8]

Did I say uranium is "depleting"? That's because the United States uses its nuclear fuel only once, disposing of the spent fuel. Technologically, however, it is possible to "recycle" spent nuclear fuel in what is, distressingly, called a breeder reactor. But reprocessing creates the opportunity for nuclear fuel to fall into terrorist hands. Because of the national security risk, the United States doesn't reprocess its spent fuel. France, however, does. To date, there have been no incidents, of which we are aware, of recycled nuclear materials falling into terrorist hands.

Because we don't recycle our spent nuclear materials, we accumulate nuclear waste. The spent fuel rods are high-level waste. That is, they'll remain dangerously radioactive for hundreds of years. Think about that. Hundreds of years, maybe longer, during which this frightening garbage will have the power to sicken or kill.

How do we safely store this dangerous detritus of our massive electricity appetite? We need to minimize the risk to ourselves and our descendents. You can't put it in buildings or any other man-made structure, because they get demolished or fall down. Even if you label the buildings and containers of waste with today's recognizable nuclear symbol—or, better yet, with the universal skull and crossbones of pirate ships—future generations may not recognize the symbol nor understand the danger.

The alternatives are to shoot the waste into outer space or bury it in geological structures deep beneath the surface of the earth. Outer space is too expensive and too risky. Do you trust the government to figure out the safest geological structures in which to put this long-lived liability? If not, you're too late: DOE long ago selected Nevada's Yucca Mountain as the nation's leak-proof, earthquake-safe repository for high-level nuclear waste.

As of 2007, however, no high-level waste has been stored there. In addition to local opposition in Nevada and lingering questions about the suitability of the geology, politicians and other watchdogs of public health and safety refuse to let high-level waste be transported through their states and cities to get to Nevada. Inasmuch as most nuclear power plants are situated in the eastern third of the country and Nevada is in the western third, we're talking about as much as 2,000 miles of transport on U.S. highways, through heartland communities.

Instead, the nation's high-level nuclear power plant waste is stored on-site in cooling ponds at power plants around the country.

Don't forget the homeland security risk. A terrorist could steal fissionable materials and make a bomb, for detonation anywhere. Atomic bomb–grade material is not needed to build a dirty bomb. Another homeland security risk is the possibility that a terrorist could fly a plane into a nuclear facility. Supposedly, the containment structure is designed and constructed to withstand such a blow. Luckily, we have not had occasion to find out.

Too cheap to meter? Not hardly.

The Balance

The nuclear power option boils down to how one assesses three factors: (1) relative cost, (2) relative risk, and (3) the waste issue.

Although the consequences of a nuclear event are likely to be catastrophic, the likelihood of such an event—as evidenced by our own experience—seems minimal. Americans accept 190,000 deaths every year from medical mistakes; 150,000 from smoking; 100,000 related to alcohol; and 50,000 on our highways. The nuclear industry calculates that its body count is fewer than two per year.[9]

If we as a society deem global climate change to be the premier risk to our planet, we will weigh the homeland security risks of nuclear to be less by comparison. One could expect that we would therefore create financial disincentives for coal-fired power plants in order to minimize the carbon risk. If these disincentives take the form of carbon taxes or financial adders of some kind, this would make nuclear less outrageously expensive by comparison.

As part of our commitment to massive power generation without carbon emissions, we could try to figure out the waste angle. Public education campaigns to explain the safety of the vessels holding the waste would need to be undertaken. Special provision would need to be made for transporting—more to the point, permitting the transport—of waste to Yucca Mountain or the Waste Isolation Pilot Plant (WIPP) that has been in operation in Carlsbad, New Mexico, unbeknownst to most Americans, for twenty years.

If we really want to encourage the growth of nuclear power, we'll need to approach design and construction much as the French do: settle on one or two designs, optimize them for safety and operational parameters, and then monitor design and construction for compliance. The way we approach each nuke (nuclear power plant) in the United States today—by designing and engineering each one individually—adds significantly to the cost. We could reduce the price tag significantly by adopting a cookie-cutter design approach.

The bottom line, though, lies in the answer to this question: Do the benefits of nuclear power outweigh the risks and costs? The verdict to date in the United States has been no.

Natural Gas

Natural gas is a mix of gases, primarily methane. It should not be confused with gasoline, which is made from petroleum. Natural gas was first used in the United States to light street lamps in Baltimore. It's come a long way since then.

Natural gas is found in subterranean reservoirs along with oil. If you find oil, you'll find natural gas. Although 85 percent of our natural gas is currently domestic, guess where most of tomorrow's natural gas is located: it's under the same five Middle Eastern countries that have the oil we want.

I say "future" supplies because our domestic production peaked back in the 1970s along with our domestic oil. Unless we curb our appetite for natural gas, we'll need to import it. Why do we hunger for natural gas? Many answers lie under the "good" column of the natural gas ledger.

The Good

Natural gas is the youngest member of the fossil fuel family. Consequently, it burns cleanest and, from a greenhouse gas perspective, puts the least carbon dioxide into the atmosphere—less than half the amount emitted by coal.[10]

Globally, natural gas is plentiful. In fact, it's so plentiful that we used to "flare" it (burn it off as waste gas) in the United States before we realized its value. To this day, it is flared in the oil fields of the Middle East. This is why some people talk of importing natural gas from that region, though it would have to be liquefied for transport purposes. Hence the acronym LNG.

The primary reason for our voracious appetite, however, lies in the versatility of natural gas. We heat some 70 percent of our new homes with it. It can also run our dryers and grill our steaks.

Almost half our natural gas use is for industrial purposes. It supplies process heat for big industries such as pulp and paper, cement and asphalt, chemicals, plastics, and even petroleum refining. We also use natural gas in the agricultural sector. It's both an energy source for food processing and a key ingredient in the fertilizer spread on farm land.

In compressed form (CNG), natural gas fuels fleet vehicles, such as city buses or government vehicles.

Finally, when used in power plants—in what is fundamentally jet engine technology—natural gas generates electricity. These are the power plants that utilities can turn on and off quickly to meet peak demand needs. These peakers are so valuable to utilities that, before natural gas prices took a dramatic upward turn in recent years, more than 180,000 of them were on order across the country.

The Bad

In less than a decade, the price has almost doubled. This is the key problem with natural gas: it costs more than the other major electricity-generating fossil fuel (coal), and it will cost even more in the future. Another problem is the volatile nature of the prices. Between 1999 and 2007, there were any number of price swings. In addition to the increasing cost of natural gas, price volatility makes it difficult for small businesses and individual consumers alike to plan and budget.

Natural gas prices always have been volatile. That is due to the nature of the industry itself and the difficulty of inventorying and storing the fuel. The umbrella term *natural gas* also includes propane, the fuel of necessity in rural and mountain areas that aren't served by natural gas pipelines. For better or worse, the propane distribution network has been characterized by small mom-and-pop businesses, not known for accurate inventories. Consequently, propane prices used to swing with seeming unpredictability, based on supply and demand pressures that are exacerbated when stored inventory is lacking.

For a while, the volatility tended to mask the underlying upward cost trend. Recent price increases have been so dramatic, however, that the trend has become obvious. This is because natural gas production in the Lower 48 peaked about 20 years ago. This doesn't mean that all our natural gas is gone, but it does mean that the cheap, easy-to-find gas is depleted. It means that we have to drill more just to produce the same amount, and it costs more to bring the gas out of the ground. Whenever you read stories in the business section of your newspaper touting the good news that the drilling rig count is up, it probably isn't cause for consumer celebration. Compare the rig count with production volume, and you'll see that we're drilling more, but production is, overall, pretty flat. Higher rig counts may be good for the drilling business, but they're probably bad for consumer prices.

Some have suggested that we could import natural gas from Canada. Well, maybe. Given Canada's need for heat and power in its own growing economy, we shouldn't be surprised if Canadians limit the amount they're willing to sell to us.

We can liquefy natural gas, which means we don't have to rely on pipelines for transport. Today, something like 3 percent of our natural gas supply is LNG. Some have suggested that we can import LNG from the Middle East, where the natural gas supply is as plentiful as the oil supply. Somehow the national security conversation gets reduced to a whisper in this scenario. At a time when we're investing in ethanol to fuel our cars as a means of getting off Middle East oil, it doesn't make sense to create a new kind of dependence on that same region.

Moreover, bringing LNG tankers into our ports poses homeland security issues. The tankers are thought of in some circles as floating bombs. Port communities have expressed some trepidation that LNG coming into the terminals could congeal and explode. Environmentalists object that pipelines connected to the port terminals would be placed in ecologically sensitive wetland areas.

The Balance

Back in the 1980s and early '90s, when renewable energy resources first started to receive serious attention as potential energy supplies, natural gas was viewed as the "bridge" fuel to the future. This was because it was thought to be in abundant supply, it was affordable, and because it can be used in modular, smaller power plants—similar to the modularity of renewable energy technologies. Most important, however, renewable energy proponents valued natural gas as a backup fuel that could be turned on and off when the sun failed to shine or the wind didn't blow—a way to "firm" the intermittent power from renewable energy resources.

Time has shown that domestic supplies of natural gas are neither abundant nor affordable. However, the value of natural gas as a relatively clean fossil fuel is greater in today's climate change–conscious world than ever before. Although natural gas will probably continue to be too expensive to be routinely burned up to generate electricity and will probably be reserved for high-value uses, such as a feedstock in the petrochemical industry, it likely will continue to meet niche, though not necessarily small, electricity needs. These would include peak power generation and blending with utility-scale renewable resources to firm the power.

Natural gas will be pricey. However, just as it can be blended with renewable resources to firm power, so too can its cost be blended, or averaged, with the cost of renewable energy to effectively lower the renewables' price. The capital costs of renewable energy technologies are declining dramatically, and the fuel is free. Consequently, packaging natural gas technologies with renewables when technologically feasible should make economic sense.

Conclusion

Twentieth-century technologies served us superbly well in the twentieth century, and they continue to perform well today. Nothing is perfect, and we had no choice but to live with the adverse consequences of coal, nuclear, and natural gas power.

Among the adverse consequences was the acid rain that caused massive fish kills in New York's Adirondack lakes in the 1960s. Toxic rain was created by emissions from coal plants in the Midwest and pumped into the jet stream through tall stacks. Transporting the emissions via the jet stream to another region was an unforeseen and unintended consequence of regulatory compliance. This was an early warning of how much our planet is shrinking, figuratively, and an indicator that actions in one region can have unintended consequences in another.

Electric utilities do their best to manage and control the risks. By and large, they succeed on a grand scale that is unprecedented in engineering experience. In addition, twentieth-century electricity technologies enjoy what Hermann Scheer dubs the "home team advantage."[11] If we gave any thought to our electricity system—though most of us don't—we probably would choose the path of least resistance and would stick with the devil we know rather than change the system.

But two things have changed. For one, some of the adverse consequences—such as accumulating greenhouse gases and resource depletion—are mounting. We as a society may be reaching our tipping point of tolerance for them. Second, we have new technology options that can be tools for change if we just reach out and use them.

It's understandable that many of us have not given much thought to what could go wrong, and it might seem that I dwell heavily on some of the drawbacks of twentieth-century technologies. However, given that the system is aging and stressed, it would be wise to be aware of potential problems and try to address them before they become crises. I want to be clear about what some of these issues are.

Electricity 2000+: Power, Today and Tomorrow

There are no silver bullets. There is only silver buckshot.
—Bill McKibben, author, educator, and environmentalist

Today we have technology options that we lacked yesterday. If we use them, we can start to mitigate the adverse consequences of yesterday's choices. More to the point, going forward we can reduce those consequences.

We can not mothball our behemoth power plants. We thrive in an electricity-hungry economy, and we're not likely to put ourselves on that kind of austere diet. In fact, the hottest new entertainment and lifestyle gadgets, as well as commercial and industrial equipment, increase our collective appetite for electricity. Even if they're energy efficient, there are more of them and, in the aggregate, they use more electricity. Think of your recent big purchases: how many of them use electricity and, even worse from your utility's perspective, create a plug load*—that is, TVs, computers, CD players, and other equipment that continue to draw electricity even after you turn them off?

Think of our homes. How many of us live in homes that are smaller than the ones in which we grew up? Not

* This is why utilities also refer to them as "phantom" or "vampire" loads.

many. The American dream calls for us to do better than our forebears, and better often means bigger. Your utility can tell you that our significantly larger footprint on the land creates a parallel appetite for electricity. Even the U.S. industrial sector has become increasingly electrified, in part for environmental compliance reasons, as electricity is cleaner than the direct combustion of fossil fuels.

All in all, U.S. electric utilities have served a 70 percent load growth in the past twenty-five years, without adding much base load generation or transmission capacity. The Edison Foundation, an arm of the electric utility industry association, predicts that demand for electricity will continue to rise in the foreseeable future, at rates ranging from 11 to 17 percent between 2006 and 2014.[1]

Although we can't scrap our twentieth-century technologies, we can supplement them with the affordable alternatives that are now commercially available. Moreover, we can continue to add new technologies and products as they become available and affordable. Distributed energy technologies, such as rooftop PV or small wind, are to today's power plants as personal computers were to the mainframes of the 1980s.

Distributed Generation

To integrate renewables into the utility system will require a very different model of electricity generation, transmission, and distribution. Incredibly, and lucky for us, it will likely be cheaper than just adding more of the same-old to today's electricity grid. Simply building onto today's grid as it is currently designed is not the most efficient solution in any case. Recall that today's grid is not, strictly speaking, one grid. It is, in fact, three regional grids that are not always connected where we need them to be, and not necessarily in sync with one another so that electricity can flow easily and where we want it to go.*

Most important for the purpose of this discussion, the grid was not designed to transmit electricity generated from renewable resources. Consequently, high voltage wires do not reach into remote sunny and windy regions. This creates stranded assets of sun and wind that can't be tapped for their electricity-generating potential. Upgrading the efficiency and adding enough high-power transmission capacity to meet our anticipated needs, if we simply build onto the current model, could cost billions for transmission additions alone.[2] Given the myriad other pressures on our national wallet, this isn't an investment that's likely to be made.

Picture, if you will, today's model: It's a large central station power plant—a nuke, a coal plant (with the associated coal piles and multiple railroad sidings), or natural gas. Transmission lines lead away from it and deliver electricity to distribution points. From there, distribution lines lead to your neighborhoods, homes, and businesses. In today's model, the electricity flows one way: from the power plant to you.

What if, however, you were to generate some of your own electricity? What if you put solar electric panels on your roof and actually fed electricity back into the line at sunny times when your system generated more electricity than you needed? If you own a commercial warehousing facility with a large flat roof, that urban roof could be a veritable field for solar electric arrays. It could become a mini power plant, generating electricity and feeding it back to the grid for consumption somewhere else in the system.

* Except for Texas and the northeast tip of New England, all of the Lower 48 fall into either the Western Interconnect or the Eastern Connect. Texas has its own grid, the Electric Reliability Council of Texas (ERCOT). The tip of New England lies in the eastern Canadian grid.

This is called distributed generation. The generating plants, whatever form they might take, are dispersed and located at or near the point of consumption. This increases transmission efficiency and would eliminate most of today's line losses in the distributed part of the system.

Now picture our utility model as described above, but add these multiple mini power plants located at the point of consumption (i.e., your home or business). These power plants could be powered by solar, wind, or both. They could also be powered by biomass. Urban biomass supplies could include landfill gas, urban wood waste, suburban lawn trimmings. In rural areas, biomass could come from plant and animal wastes and perhaps even dedicated fuel supply crops such as switchgrass, drought resistant prairie grasses, or fast-growing poplar trees. Large facilities such as hospitals, schools, or industries could install fuel cells and create microgrids in building complexes or neighborhoods.

These mini power plants could take any form and would be distributed throughout the system. The end result is that they would take a "load" off the existing generation and transmission facilities. In the aggregate, they're likely to reduce the load enough that many (if not all) additional central station power plants would not need to be built. These distributed resources could also use local distribution lines and therefore could reduce the need to add expensive transmission capacity.

Maybe a lot of these distributed mini power plants would not be big enough to generate excess electricity. Maybe they would succeed mostly in reducing the need for electricity from the central station power plant. Maybe you would add energy efficiency to your personal mix, reducing the need for electricity even further. You might not have built a mini power plant, but you would succeed in reducing—and, in some service areas, perhaps eliminating—the need for an added nuclear or coal plant.

In fact, one utility combined distributed renewable energy sources and aggressive energy efficiency to make up for the loss of its nuclear power plant in the 1990s. The Sacramento Municipal Utility District (SMUD) took a planned nuke off the planning books after the citizens directed it to do so in a referendum. Within five years of the referendum, SMUD installed 2 megawatts (MW)* of PV on-site at the mothballed nuke, 5 MW of wind, 134 MW of geothermal electricity, and 4 MW of PV distributed on residential rooftops.

SMUD's rooftop PV program provides an interesting model for other utilities. The utility owned the PV arrays and used the homeowners' roofs, similar to leasing land for wind farms. Utilities, with the obligation to serve, are understandably nervous about turning over "the keys to the car" in terms of letting other generators onto the system, whether they are industrial cogenerators or dispersed small generators. SMUD retained ownership of the equipment and controlled operation of the system.

In addition, SMUD was able to reduce its peak load by 12 percent through energy efficiency. It planted 300,000 shade trees, reducing indoor cooling requirements as much as 40 percent; it helped customers purchase more than 42,000 superefficient refrigerators; and it provided rebates for cool roofs (rooftops with sun reflective coating). SMUD spent 8 percent of its gross revenues on energy efficiency and succeeded in holding rates constant for ten years. In contrast, rates would have skyrocketed 80 percent had the utility completed construction of the nuclear power plant.

* A megawatt equals 1,000 kilowatts—enough electricity to power 750 to 1,000 homes. Power plants range in size from 350 to 750 or more than 1,000 MW.

Buildings

We can't talk about a new electricity model without specifically addressing electricity use in buildings. Seventy-one percent of the electricity consumed in this country (and 40 percent of primary energy) is in buildings. The 81 million buildings in the United States comprise the largest single energy-consuming sector in our nation's economy.[3] This is the consequence of a building's orientation toward the sun, its architectural design, the tightness and insulation of the exterior shell, insulation level, quality of windows, general quality of construction, and—increasingly—the electricity appetite of equipment inside the building. With regard to how we use that equipment, especially heating, ventilation, and air conditioning (HVAC), we "operate" buildings.

Start at a building's beginning. First, orient it properly on the lot in order to take full advantage of the heating and natural daylighting properties of our solar system's power plant. Minimize the amount of window space on the east and west sides, because this is where the building soaks up unwanted solar heat gain and glare in the morning and afternoon. Developers and municipal planners should plot streets so that as many of the structures as possible can face south.

Next, design the building to maximize the sun's benefits. This is passive solar design. Size the roof overhangs so that the summer sun doesn't overheat interior spaces, especially in warm climates. Install windows with glazing to minimize glare and heat gain. Design your window space to optimize the sun's natural daylighting properties without paying the penalty of unwanted heat gain. Add the appropriate level of insulation for the climate zone. Make sure the building is well constructed and without leaks.

Top it with a cool roof to reduce solar heat gain. Until

recently, that meant painting the roof white, to create an albedo effect and reflect heat just as arctic glaciers do. Aesthetically, that worked okay for flat commercial roofs, but maybe not so much for homes. Now, however, one can purchase "cool roofs" in a variety of colors.

A building that is oriented toward the south and that includes the foregoing passive solar design features uses some 60 percent less energy than conventional structures. That percentage goes up even higher if you install energy efficient appliances and add rooftop solar water heating and electricity. These days, PV can be integrated into properly oriented and sloped roofs, or even the sides of buildings, as an alternative (albeit an expensive one) to conventional PV arrays.

Now we have what is called a net zero, or near zero, energy building. Futuristic? The future is now. Several production builders in California have done this already. Shea Homes, for example, offered near-zero energy homes in one-third of its upscale Scripps Highlands development.[4] Production home construction is a market that can absorb the added cost of PV: builders can reap the cost savings of bulk power purchases and home buyers can put the added cost on their mortgage. Moreover, builders can contract with PV suppliers to provide turnkey installation services.

DOE's National Renewable Energy Laboratory (NREL) and Denver's Habitat for Humanity volunteers built a demonstration zero energy home (ZEH) in 2006. The home features superinsulated walls, floors, and ceilings; efficient appliances; compact fluorescent lights; solar water heating; a 4-kilowatt rooftop PV system; and windows with a special coating to reduce heat transfer (windows are always a problem from an insulation standpoint). The home is inhabited and its energy consumption is monitored. So far, it seems to be on track as a true ZEH. Unfortunately, it was expensive to construct, primarily because of the PV and the special windows. Costs will need to come down before this ZEH model can be marketed on a large scale, especially to those who can't afford initial high capital costs any more than they can afford high energy costs.

By offsetting the need for the electricity that a power plant would have had to generate, transmit, and distribute to these buildings, these homes have effectively created a demand-side resource of electricity. Imagine the impact of structures like this across a utility service area. Although it did not envision zero energy buildings, one utility in particular aggregated energy efficiency improvements across its service territory and benefited enormously by doing so.

Virtual Power Plants

During the 1990s, the municipally owned utility in Austin, Texas, "constructed" an energy efficiency "power plant."* As it planned to meet the needs of its growing population and economy, Austin decided to put its money not into the bricks and boilers of a conventional power plant, but to invest instead in programs that resulted in energy efficiency throughout its utility service territory.

This is a good time to differentiate between energy conservation and energy efficiency. Conservation is a wonderful thing. It's rooted in human behavior, such as turning off lights, turning down the thermostat, disconnecting plug loads, and so forth. During California's natural gas crisis of 2000, energy conservation is what saved the system. Within twenty days, Californians voluntarily reduced their electricity consumption enough to avert major blackouts that otherwise would have occurred.

* Austin actually called it a "conservation" power plant, as this effort commenced before the term *energy efficiency* became popular.

But let's face it: humans are unreliable. We often forget to turn off lights, unplug the computer and TV, and lower the thermostat. Customarily, a utility can't rely on our behavior to reduce its load requirements in perfectly predictable fashion over a sustained length of time.

In contrast, a utility can rely on energy efficient technologies that, once installed, continue to work and reduce load. These would include high-efficiency lights and appliances, added insulation, passive solar design in new buildings—all of these and more.

This is what Austin did. It created programs to encourage investment in energy efficiency. This included energy efficient building codes that were actually enforced. (Many jurisdictions do not effectively enforce their codes, usually for budget reasons, as enforcement is labor intensive). Austin also underwrote rebates for high-efficiency equipment and appliances, and instituted other programs. The utility kept track of these investments and did the engineering calculations to ascertain how much electricity was offset when the equipment was installed. The monitoring and tracking processes were more complicated than I make them sound, but this was the gist of it.

After about a dozen years—which, coincidentally, is about how long it would take to acquire the permits and construct a coal-fired power plant—Austin had booked 550 MW of affordable and sustained energy efficiency on its system. Because it had "constructed" this virtual power plant, Austin took a 450 MW coal-fired power plant off its planning books. This was during a time when the local population almost doubled and the local economy grew by about 46 percent.

Part of Austin's economic growth may well have been related to its investment in energy efficiency. Recall our discussion about the labor-intensity of energy efficiency and the local multiplier effect. That is, unlike central station power plants that are likely located somewhere else, energy efficiency requires local labor for retailing, distribution, installation, and other functions. Money spent on those salaries is likely to be spent locally, and then respent again, creating a multiplier effect. The multiplier for energy efficiency in Osage, Iowa, was $2.23, in comparison to $1.66 for a central station power plant.

In addition, both Osage and Austin found that energy efficiency is cheaper than constructing new power plants. This enabled them to attract desirable new industries, lured by the promise of affordable electricity for business operations and enhanced quality of life for their employees.

How do we know energy efficiency is cheaper than new power plants? Regulated electric utilities across the country are required, and have been for almost twenty years, to create and carry out energy efficiency programs of one kind or another. One investor-owned utility, headquartered in the Midwest and serving multiple states, recently reported that its energy efficiency programs (including rebates) "produced" energy efficiency at a cost of less than 3 cents per kilowatt-hour.

Austin's virtual power plant was built exclusively from cost-effective energy efficiency, and the energy savings achieved from any one measure were small. Yet, when aggregated across the utility's service area, the savings totaled the equivalent of a power plant. It didn't take any longer to accumulate these massive savings than it would have taken to construct the coal-fired plant they took off the books.

How many similar or larger plants could be built across the country? The Alliance to Save Energy, a nonprofit organization dedicated to the advancement of energy efficiency, looked at this issue a couple of years ago. It examined a suite of four measures and policies and concluded that, in the aggregate, they would equate to 557 power plants. Those policies included increased appliance-efficiency standards, efficiency standards for commercial air conditioning, energy efficient design and construction of new buildings, and efficiency upgrades in existing building stock.

Austin "constructed" its virtual power plant before rooftop solar electric had become cost-effective or commercially available on any meaningful scale. Imagine what the savings of conventional electricity-generating fuels would be if distributed renewable energy technologies were added to the mix and a "green" virtual power plant were constructed.

Following are descriptions of the technologies that could be incorporated in a virtual green power plant. The discussion is organized according to whether the technologies are building-based—i.e., distributed—or utility-scale. Utility-scale renewables can be situated near existing power plants to supplement (repower) the output.

Building-Based Power

Energy Efficiency

> *Every watt not used is a watt that doesn't have to be produced, processed, or stored.*
>
> —*Richard Perez,* Home Power *magazine*

Dubbed "the fifth fuel" by a utility executive, energy efficiency is about equipment, measures, and practices, not behavior.[5] Examples include the following: CFLs, high-efficiency air conditioners, variable speed motors, optimum insulation levels, high-performance windows, and the like. These are not electricity-generating products, but in the aggregate they can offset the need for a significant amount of electricity, totaling hundreds of power plants.

Key to our twenty-first-century electricity model is electric utilities' changing their corporate self-concept. They can provide energy services, not just electrons.

The Good

When you save energy, you save more than the amount saved at the point of use. You also save the energy that would have been used to generate the electricity. Some 25 percent of our domestic fossil fuel consumption occurs in the production and delivery of energy itself.* You also save all the electricity that vaporizes during transmission and delivery. In this sense, energy efficiency is actually an energy resource—a demand-side resource.

A common example of the power of energy efficiency can be found in most homes in the United States: the common household incandescent lightbulb. Ninety percent of the electrical energy that goes into the bulb is thrown off as waste heat. Only 10 percent actually lights the light. Until CFLs came into the marketplace in recent years, lightbulbs were essentially the same technology as invented by Thomas Edison and his colleagues in the nineteenth century. Despite all the other technological innovations we've seen over the years, lightbulbs remained pretty much unchanged until recently.

If it's summertime or if you live in a warm climate, you probably pay to cool the interior of your home or business. Consequently, your lightbulb is costing you even more: it's adding, perhaps significantly, to your cooling load and

* This includes the energy used in the production and delivery of transportation fuels as well as electricity.

associated costs. CFLs, in contrast, have a higher price tag, but they use about three-quarters less electricity than incandescents, they last seven times longer, they produce a higher quality light, and they're better for the environment.

CFLs repay your initially higher cost through energy savings, probably within a year. This illuminates a really important concept: life cycle cost-benefit analysis. When making purchasing decisions about energy-using products, one should always factor in the cost of the energy needed to operate it for the anticipated lifetime of the equipment. The purchase price (first cost) of energy efficient equipment might be higher than standard models. But you'll save sufficient money in reduced energy bills to pay back the added purchase price. You have to decide if the payback comes fast enough to make it cost-effective by your standards.

Here's an example. Compare a CFL to a standard incandescent bulb:*

	Incandescent	CFL
First Cost	$0.75	$2.00
Lifetime Operating Cost	$77	$18
Lifetime CO_2 Emissions	2,800 lbs	660 lbs[6]

You might not think that the energy savings from a lightbulb are very impressive. Even the energy savings from all the lightbulbs in your house might not seem to produce dramatic savings in any one utility bill. But how about over the course of a year? How about if those savings are multiplied across many homes all across the country? It all adds up, and it gets us back to the notion of virtual green power

* Purchase price will vary by location and brand, as well as over time. The differential between incandescent and CFL remains roughly the same despite these variations.

plants constructed all across the United States.

The really good news about energy efficiency, aside from its affordability, is its tremendous upside potential. This is the direct consequence of our wasteful energy habits to date. It's not unreasonable to expect, as a rule of thumb, at least 20 percent savings from energy efficiency retrofits alone. Imagine the 60 percent potential if we design and construct buildings for maximum potential energy efficiency.

The Bad

The many energy efficient products and practices represent different technologies and expertise, and they're not organized as one energy efficiency industry. This creates an incredibly fragmented "voice" for energy efficiency, exacerbated by the fragmented nature of the residential construction industry, in which workers are often hired on a day-to-day basis. Quality control is difficult to achieve day in and day out, and is even more so if new measures and practices are introduced on-site.

In addition, builders and developers have the perception that energy efficiency costs more. This used to be true. Over the years, however, costs have come down. Today, added first costs for energy efficient design and construction amount to 2 percent or less—recouped within a few short years (sometimes months) through energy savings.

Nevertheless, perception tends to become reality, and builders and developers resist what they believe to be unnecessarily expensive building practices. In addition, it's difficult to change building practices, because so many different products, people, and actions are involved at each stage of construction. This requires diligent, sustained oversight of daily construction operations—very difficult to do in the real world.

Added first costs also create a deterrent for home

buyers. If given the choice between adding granite countertops or energy efficient windows and high-efficiency appliances to the mortgage, the homebuyer usually chooses the granite countertops.

Why are consumers making economic decisions that are clearly not in their financial interest with regard to energy efficiency? Two reasons: (1) it's not clear to them that they're making financially suboptimal decisions, because they lack information about their choices and the consequences; and (2) energy efficiency improvements are not as sexy as, say, the aforementioned granite countertops or even rooftop solar, for that matter. There is nothing obviously cool—or even obvious, in most cases—about energy efficiency.

This suggests the need for greatly improved consumer education. It also suggests the need to redefine our notion of beauty in our homes and businesses.

The Balance

Whether you're trying to construct a virtual green power plant or making purchase decisions for your home or business, always consider energy efficiency first. First and foremost, this is because the low-hanging fruit of the efficiency tree is customarily cheaper than any others. After an energy efficiency measure is installed, it continues to save money on energy bills. With the money saved, one can accumulate funds to purchase additional energy-efficient equipment or save for a renewable energy investment such as PV.

Efficiency is also important to undertake first because it enables one to downsize other energy-using equipment and reap the resulting cost savings. For example, builders can install smaller HVAC systems in energy efficient structures. They know that the building will retain the air the occupant has paid to condition (whether heated or cooled) if it has optimum levels of insulation in the walls, if the windows are energy efficient, and if leaks are sealed. A smaller HVAC system costs less, so this enables the builder to reduce the first costs accordingly. The homeowner also reaps the benefit of the reduced operating costs of a smaller HVAC unit.

Similarly, energy efficiency permits downsizing of renewable energy technologies. Even the most committed solar researchers or advocates always advise consumers to invest in efficiency first, in order to minimize the size and associated cost of a solar system.

Finally, energy efficiency makes investment in renewables such as PV more affordable. If one bundles efficiency with renewables and averages the costs, the higher-priced equipment becomes more affordable. Just like cost averaging stock market purchases.

Solar in Buildings

> *Each hour enough sunlight strikes the earth to meet the world's energy needs for an entire year.*
>
> *—National Renewable Energy Laboratory*

If only we could figure out how to harness the power of this incredible resource and do it affordably. Actually, though, due to thirty years of research and development at the U.S. Department of Energy's National Renewable Energy Laboratory (NREL) in Colorado—not to mention the numerous other R & D efforts at other national labs, colleges, and universities and in the private sector—the cost of solar technologies has come down by orders of magnitude since the mid-1970s.

In addition to natural daylighting and passive solar

design, there are several solar technologies for application in buildings: solar thermal (for water and space heating), solar electric (PV), and transpired solar air collectors.

Solar Water Heating

Like passive solar design, solar water heating is often viewed as a demand-side resource because it offsets the need for conventional fuels or electricity to heat water. There's nothing new about this rooftop technology, though it has been improved over the years. Many Americans are familiar with it from its introduction into the U.S. market—complete with federal tax credits—in the late '70s and early '80s, though some roofs sported solar water heating as long ago as 1911.

The technology is straightforward. Dark-colored solar collectors (panels) are placed facing south. A heat-transfer fluid runs through coils, absorbing the sun's heat. In most systems, the heated fluid is pumped into the building, where it's stored in a tank until needed. Most systems are active; that is, they employ pumps to move the heated fluid.

The Good

Heated water for dishwashing, laundry, and bathing is the second biggest home energy cost. (Space conditioning—heating and cooling—is first.) Solar water heating usually supplements rather than supplants a home's conventional water heating needs. The savings of conventional energy or electricity are not trivial—50 to 80 percent of water heating costs.

If you have large hot water needs, or if water heating costs you a lot, this technology can be expected to pay for itself in energy savings in a matter of years. Especially in commercial or institutional settings with high-volume hot water needs (for example, laundry operations in prisons and hotels), this technology can be a very cost-effective investment. In commercial or institutional settings where the facility has a flat roof, building orientation toward the sun is not an issue. The panels can be mounted on frames and oriented toward the sun. The technology has been improved in recent years, mostly to make it more attractive or blend with roof lines. It's straightforward and requires minimal maintenance.

The Bad

Thirty years later, the industry still has a bad reputation in some circles, based on the poor performance of some systems and some so-called solar companies back in the 1970s. The book on how *not* to structure tax credits or financial incentives was written with the federal tax credits for solar water heating enacted by Congress in the 1970s. In a nutshell, the tax credits attracted fly-by-night operators to the field, leaving many customers stranded with nonworking systems, a second lien on their homes, and no easy recourse. The fly-by-nights flew, and consumers were left with "orphan" systems. Many of today's solar water heating business owners spend some amount of time and effort servicing these systems or pulling them off roofs.

In addition, a lot of the early systems were just plain ugly—eyesores propped up in looming and uneasy fashion on your neighbor's roof.

Solar water heating is still on the expensive side. Prices vary, depending on system size, retailer, and other factors. Payback will vary depending on the aforementioned, as well as how expensive your conventional water heating costs are. While it's not uncommon for solar water heating to have a six-year payback or less, this might still be beyond the investment horizon of many families, institutions, and businesses.

The Balance

Whether or not solar water heating makes sense for your home or business depends on how you answer the following questions:

- How much of your energy or electricity bill is for water heating?
- How much hot water do you use, and how much do you expect to use in the future? (Will you still live in the same house when your seven-year-old grows into a teenager?)
- Is your roof oriented toward the south for unobtrusive placement of the arrays? If not, do you have some other way to orient them properly without creating an eyesore for your neighbors?
- How long is your investment horizon? (Can you wait for the system to pay back through utility bill savings?)
- Does solar have value to you for other reasons, so that payback is not your only investment criterion?
- How much do you foresee your utility bills rising in the coming years? If they are going up, your investment may pay back sooner than you expect.

Transpired Solar Air Collectors

I'm willing to bet you've never heard of this one. This straightforward, cost-effective technology was invented about fifteen years ago.

In a nutshell, solar air collectors are sheets of corrugated metal, painted a dark color, punctured with small holes strategically placed for optimum air penetration and movement, and affixed to the south-facing exterior wall of a structure, leaving an airspace between the metal and the building. Sunlight is absorbed by the dark color, and the air is heated as it passes through the collector wall. Small fans, as well as the natural convection of heated air, suck the heated air into the building (hence the term *transpire*). On a cold winter day in Denver, this air is not warm enough to heat the interior space, but it at least *pre*heats the interior air, thus reducing heating costs.

In cool climates with intense sun, this technology might be applicable for crop drying. It's more commonly used and particularly well suited, however, for application in places such as warehouses and loading docks in cool climates. At the FedEx regional distribution facility in Denver, for example, the south-facing exterior wall is painted FedEx's signature burnt orange rather than black, costing a 10 percent penalty in energy collection.

The Good

This is a simple technology that requires little or no maintenance and costs the equivalent of 2 cents per kWh. Depending on your heating bills, this could pay for itself in as few as six months.

The Bad

Transpired solar collectors don't heat, they preheat. Their usefulness seems to be limited to commercial and industrial facilities and perhaps grain drying in selected locations. Early versions were painted a very dark color, which might have posed some aesthetic issues. However, in recent years I've seen two buildings in which the facility owner used its own trademark color or one that blended in with the rest of the structure, camouflaging the air collector. The most recent example is a Wal-Mart store constructed as an "experimental" facility in Aurora, Colorado. The air collector is painted a deep taupe and softly contrasts with the rest of the earth-toned exterior. NREL researchers estimate the energy performance penalty to be about 20 percent due to the relatively lighter color.

The Balance

If you have a suitable application in a suitable climate, plus suitable building orientation, do the math.

Solar Electric (Photovoltaics)

The term *photovoltaic* comes from the Greek word *photo,* meaning "light," and *voltaic,* meaning "power." PV cells are made from semiconductor materials like those in computer chips. When sunlight strikes these materials, the sun's photons release electrons in the PV cells. This commences a chain reaction of electrons moving between positive and negative poles that, in turn, produces direct-current (DC) electricity. Solar cells that measure four square inches generate 1 watt of electricity. This would be enough to run your watch but not enough to power your radio.

Some forty cells are connected to form a module. A module generates enough electricity to illuminate a light-bulb (hopefully a CFL). Ten modules can be mounted on an array. Ten to twenty arrays can power a household.

Since our electric utilities and electricity-using devices use alternating current (AC), most uses of PV require an inverter to convert the DC electricity to AC. The inverter and other equipment associated with the PV array are called the balance of system.

The cost of PV has come down significantly since the 1960s, when the only application of this technology was to power the electricity needs of satellites and spacecraft while they drifted in outer space. Those early arrays were assembled by hand, and the cost was as out of sight as the spacecraft they powered. These days, you see PV everywhere, even though you might not be aware of it—from handheld calculators and children's games to school crossing and highway warning signs.

Customarily, PV is used in distributed fashion. It may be grid-connected—for example, on rooftops, providing some or all of the building's electricity needs—or not. But it is often used in cases where it would be too expensive to connect to the grid, such as in remote locations, where it can cost anywhere from $10,000 to $100,000 per mile to string wire, depending on the terrain and the utility's price structure.

Applications may be residential, commercial, institutional, or industrial. Some of the larger rooftop installations include the Moscone Convention Center in San Francisco, the Mauna Lani Resort in Hawaii, and Google's corporate headquarters. Because of improved solar cell efficiencies that permit the capture of the sun's energy from broader bands of the spectrum, PV can be used in locations not customarily thought of as particularly sunny.

Sometimes PV is the cheapest alternative in urban applications, especially if it would be expensive to connect to a nearby pole. This was the case on one of the most manifestly overlit streets in America—the Las Vegas Strip. The City added a bus stop and covered it with PV for security lighting. It would have been more expensive to install conventional lighting and trench under the Strip to the utility pole across the street. An added bonus was the shade provided by the arrays.

The Good

PV is the keystone of the distributed electricity generation model. In 2007, most commercially available collectors are made from silicon, one of the earth's most abundant elements. In recent years, a shortage of processed silicon was created due to dramatically increased demand. New processing facilities are being added worldwide, so this is turning out to be a short-term issue. And industry and government researchers continue to work on improving the efficiency of individual solar cell performance in converting sunlight to electricity. This will bring down the cost, as will improved manufacturing processes.

In addition, scientists are researching other materials from which to manufacture PV. Collectively known as thin films (thinner than a flake of paint), this second generation of PV is deposited on a substrate of some kind—from glass to asphalt shingles—and is suitable for integrating into the side of a building or on rooftops. Thin film PV and roofing materials are commercially available today, but they're more expensive than conventional materials and less efficient than silicon cells.

PV is modular, low maintenance, and has no moving parts. It comes with a warranty for twenty years. The fuel that powers it is free. Moreover, the sun shines brightest, generating the most electricity, during some of the same hours of the day when air conditioning needs are greatest. In utility-speak, that means that peak solar is roughly coincident with peak demand.

A recent study conducted at Scripps Highlands revealed that PV added to the resale value of homes in an otherwise conventional upscale neighborhood. Many of the first owners of the homes bought the property not for the PV, but because it made sense for other reasons. Over time, however, they came to appreciate the lower utility bills and the PV that made that possible. This turned out to be a selling feature when they put their homes on the market.

The Bad

Cost is the number one issue with PV. This technology was not remotely affordable until the end of the twentieth century. Now, despite dramatic improvements in cell efficiency and manufacturing processes, PV is still expensive.

First-generation, silicon-based PV costs between 18 and 30 cents per kWh. This is, in large part, a function of solar cell efficiencies—which are still low, even by comparison to a coal plant. Commercially available silicon solar cells are about 15 to 22 percent efficient, and commercially available thin film technologies run around 9 percent.*

In new homes, PV adds to first costs. This is why PV proponents prefer to talk about value rather than cost. Sure, the bottom line is important. But don't forget: Americans purchase bottled water, and they drop in at coffee shops on the way to work to pick up designer lattes that cost $4 or more a pop. Price alone doesn't drive our purchasing decisions.

The other key issue with PV is that solar is an intermittent energy resource. Quite simply, the sun doesn't always shine. When the sun doesn't shine, solar cells don't generate electricity. Battery backup provides reliability in case the grid goes down, and, in fact, this might be a very smart investment for commercial operations with 99.9999 percent reliability demands. But it doubles the price of the system and adds the standard issues that go with batteries (shorter life span than the PV array itself and disposal issues, to name two). Residential PV owners seldom invest in their own battery backup.

Alternatively, PV owners can use the electricity wires for storage—i.e., to connect to the grid and use conventional electricity much of the time, especially on cloudy days and at night. This is the model most residential PV owners choose.

Remember the Demand Café? Your utility "chef" will

* It has been estimated that if PV cells were 100 percent efficient—though no generating technology is—a PV array measuring about five feet by nine feet could power an entire home in Phoenix.

serve you, but using the utility's wires as backup for your PV system is sort of like bringing your own lunch to the café, occupying a big corner booth, and ordering drinks to accompany your homemade sandwich. From the twentieth-century utility's standpoint, this is a little cheeky on your part. But just as manners and mores change over time, so too is electricity "etiquette." In our twenty-first-century distributed generation model, this will be standard procedure—like the family cafeteria at a ski slope.

For now, your utility chef will permit you to connect to the grid and will not give you surly looks if you pay interconnection fees, agree to let the utility disconnect you from the grid if needed, and assent to whatever conditions the utility imposes. All this will cost money, which brings us back to the number one negative with building-integrated PV: cost.

The Balance

As research and development continue to improve the efficiencies of PV and as manufacturing techniques improve, the cost will come down. At the same time, the cost of conventional generating fuels will continue to rise. At some point in the near-to-mid-term, the cost curves will cross. This might happen sooner if a carbon tax or other penalty for environmental degradation is levied on fossil fuels. It will also happen sooner if government-funded PV research picks up the pace. The Bush Administration has set a goal of making PV cost-competitive with conventional electricity by the year 2015.[7]

Small Wind

We hear a lot about utility-scale wind, but small wind (10 kWh or less) can also be quite cost-effective. It can be either grid-connected and placed on buildings, or off-grid, to offset the need to string wires to a remote location. Humans have used wind for hundreds, if not thousands, of years. Before electricity, windmills captured the power of the wind to turn a shaft that turned a millstone to grind grain or drive a pump to lift water.

We don't call them windmills anymore. They're wind turbines. They have three blades mounted on a shaft, forming the rotor. The propeller-like blades act like airplane wings. When the wind blows over a blade, a pocket of low-pressure air forms on the downwind side. The low-pressure pocket pulls the blade toward it (lift), causing the rotor to turn. The lift is actually stronger than the wind blowing against the front side of the blade (drag). The combination of lift and drag cause the rotor to spin. The turning shaft spins a generator located at the top of the tower (in a box called the nacelle) and makes electricity.

The Good

If you live in a windy area, and especially if you need electricity in remote locations not served by the grid, small wind is for you. Stand-alone small wind turbines are especially useful for pumping water and for communications. In remote areas of China, small wind and PV are often used in tandem. Grid-connected small wind can also help cut electric bills.

New models that are particularly suited for rooftop installation recently hit the market. Residential and commercial customers alike may now take advantage of "urban" wind.

The Bad

The power of prevailing winds is a direct function of distance off the ground. The higher you go, the better the wind resource. This is an issue if you mount a small turbine on an urban or suburban roof. If the tower should fall down, there's greater risk of injury or damage in a populous area than there is in a sparsely populated place. In urban and suburban applications, building codes and homeowner association covenants are likely to prohibit small wind on buildings. In addition, insurers raise liability issues. Consequently, the urban and suburban residential and commercial applications may be limited. Aesthetics may also be an issue for some.

It goes without saying that the wind doesn't always blow. The intermittent nature of the resource is another negative, though this is not as critical a factor as with utility-scale wind turbines that must be integrated into a generating system that needs to meet a 99.9999 percent reliability standard.

The Balance

On balance, rural applications of small wind are probably easiest to achieve at present. This may change over time. Rural residents are accustomed to the look of wind turbines through their history with windmills (some of which are still standing). In addition, building codes and covenants are not as likely to present barriers in rural areas as they are in urban and suburban communities.

Hopefully in the near future wind manufacturers will return to constructing midsized turbines, suitable for generating electricity at the community scale and transmitting it over local distribution lines. These turbines range in size from 10 to 250 kilowatt capacity. They don't require large transmission lines, so they can minimize issues associated with bulk power transmission. At the same time, they're large enough so that several machines can help electrify neighborhoods and small communities. They're especially well suited for windy rural communities.

Now that we've covered the renewable resources of the air and sky, let's go underground.

Geothermal Heat Pumps

Geothermal means "earth heat." Geothermal energy comes from underground and constitutes a vast resource. To date, however, geothermal has been the Rodney Dangerfield of energy resources: it gets no respect. That might be changing, mostly due to pioneering efforts by a handful of utilities that have tapped into this resource and to new scientific and engineering assessments of the resource.

Geothermal heat pumps (GHPs), also known as ground source heat pumps, take advantage of the relatively constant 50-degree temperature about six feet underground. They move the heat from the earth (or a groundwater source) into the home in the winter and pull the heat from the house in the summer, discharging it into the underground piping loops. If you add a water-heating dimension, it will use this discharged warmth to heat water in the summer.

The Good

GHPs are among the most efficient and cost-effective ways to heat and cool a home and to provide hot water. Because they have fewer mechanical parts, and because those parts are insulated and protected underground, they're durable. They also operate quietly, because there are no outside condensing units such as conventional air conditioning systems. They operate effectively in all major climate zones, especially those with high humidity, performing best where summer and winter loads are balanced.

GHPs in both new and existing homes can reduce energy consumption by 25 to 75 percent (depending on the system and to what you're comparing it), relative to conventional and older systems.

The Bad

Installation involves digging, on average, six or eight feet down. This is expensive and can pose some element of risk, depending on the soils. This is not for the do-it-yourselfer. Correct design and construction, as well as pressure testing, are an absolute must.* Underground leakage is a potential hazard. Trenches or shafts must be dug in which to lay the underground pipe loops. For retrofits, this means digging up part of the yard, with all the mess, noise, and hassle that entails.

There is also the issue of cost. Although GHPs pay for themselves in energy savings over time, the equipment and installation (which is nontrivial, given the digging required) entail significant up-front costs.

The Balance

Given the efficiency and other benefits of GHPs, they make a lot of sense in new construction, when it's easier to dig down and lay pipes. GHPs make so much sense in some regions that some utility companies run programs that make it as easy as possible to retrofit with GHPs.†

Utility-Scale Electricity Generation

I can't resist saying a word about energy efficiency as it applies to utilities themselves. Our entire electricity generation and delivery system is aged and inefficient. Recall that about two-thirds of the coal inputs to a power

* Actually, none of these technologies are for the do-it-yourselfer. Solar water heating requires the expertise of a plumber or certified installer. PV and grid-connected small wind require an electrician or certified installer.

† One such utility is the Delta-Montrose Electric Association, a rural co-op renowned for the pioneering GHP program it has run successfully for years in Colorado.

plant are thrown off as waste heat and toxic emissions. U.S. utilities waste enough energy every year to meet the power needs of Japan. There is considerable upside potential for improved energy efficiency in conventional power plant facilities and operations. Improved transmission lines—say, through superconducting materials that lower resistance and resulting line losses—have efficiency potential, as do improved transformers and other technology modernization in the utilities' own systems.

Following are overviews of renewable energy technologies that utilities can integrate into their electricity-generating portfolios.

Utility-Scale Solar Power

> *Nobody in their right mind should be building a coal plant. ... I'm almost convinced that the cheapest plant would be solar thermal.*
>
> —*Vinod Khosla, founder, Sun Microsystems, and legendary venture capitalist*

Sunlight and mirrors. This is what Concentrating Solar Power (CSP) is all about. The term CSP actually refers to a family of utility-scale solar technologies: concentrating PV, parabolic troughs, central receiver "power towers," and dish/Stirling engine natural gas hybrids. Regular PV can also be employed at utility scale.

It would get too technical too fast to describe each technology. Suffice it to say that what they all have in common is that, essentially, they use mirrors to concentrate the power of the sun—not unlike a magnifying glass—and thus intensify the heat to create steam and run a generator. At heart, this is fundamentally the same approach as in coal and nuclear power plants: it comes down to heating something to turn a generator shaft and generate electricity.

The other thing the solar technologies have in common is that they require excellent sunlight (insolation) and relatively clear air so that the sunlight is not diffused by particles of dust, pollution, or moisture. They also need lots of land on which to place the collectors and mirrors, and they require access to major electricity transmission lines.

You probably haven't heard a lot about this technology, but it has been around—and generating electricity—for a long time. A facility in Barstow, California, generated electricity from a central receiver power tower for about twenty years. The 300-foot centerpiece tower was surrounded by 1,800 mirrors that reflected and concentrated sunlight onto the central tower. The key to this technology is the fluid used in the power tower to create steam. In its first version, Solar One, it used water. The next version, Solar Two, used molten salt. The properties of molten salt permitted it to retain heat for some time after the sun sets. This plant was capable of generating sufficient electricity to power about 10,000 homes.

Several parabolic trough projects in California's Mojave Desert supply more than 305 MW of generating capacity to Southern California Edison. This is the equivalent of a small coal plant. Arizona Public Service runs a 1 MW parabolic trough generator, and a 64 MW plant came on line in Boulder City, Nevada, in 2007. Eight MW of PV, augmented by some CSP (concentrating PV), is scheduled to come on line in Colorado's San Luis Valley in 2008. This utility-scale technology is finally coming into its own and is benefiting from the impetus provided by the Western Governors' Association goal of installing 1,000 MW of CSP as part of its larger clean energy initiative.

The Good

As it happens, the ideal solar resource for CSP technology is concentrated in the region where the population is growing most rapidly. The desert Southwest receives the intense solar radiation and enjoys the wide expanses of flat, unshaded land required for CSP. This region is also home to fifteen of America's twenty fastest growing metropolitan areas. Just as the sun shines brightest during the time of day when electricity demand tends to be greatest, it also shines most intensely on this region with the greatest rate of population increase and associated load growth.

The solar resource in the region is astounding. NREL once calculated that, if the 100 square miles that constitute the Nevada Test Site—where atomic bombs were detonated during Cold War days—were covered with PV arrays, the amount of electricity generated could meet the needs of the entire United States. Talk about turning swords into plowshares …

There's good news regarding cost. In 2007, CSP costs about 13 to 17 cents per kWh. This is steep for most of the country, but not outrageously so. Moreover, the future looks even brighter for CSP: DOE's research goal is to achieve 7 to 10 cents per kWh by 2015 and 5 to 7 cents by 2020. Even if these specific cost goals are not reached, CSP promises to be cost-competitive with "conventional" technologies within those time frames, as the costs of fossil fuels and uranium continue to rise.

The Bad

It's the flip side of the The Good. The resource base and suitable topography exist primarily in the Southwest. Although the resource is abundant enough to meet the needs of the entire country, the immediate question is how to get the solar electrons to the needy markets outside the region. A lot of costly high-voltage wires would need to be installed to carry these electrons to the rest of the country. This is the issue that NREL analysts encountered when they did those calculations related to the Nevada Test Site. There's nothing trivial about building needed transmission. Not only is it costly, but it also takes years to acquire needed rights-of-way.

Concentrating solar technologies also require water for cooling, which may be an issue in the desert regions where the insolation is best for these technologies. This is why Xcel Energy's solar facility in Colorado's arid San Luis Valley consists mostly of regular PV, which doesn't have the water cooling requirements of CSP.

The Balance

CSP is less expensive than rooftop PV and not much more expensive than nuclear (not counting nuclear's externalities, of course). If a carbon tax were enacted and the price of coal were to rise due to policy, market forces, and transportation costs, CSP could be cost competitive—even without monetizing its external benefits—within a few years. Although the utility-scale solar resource is limited geographically, the power potential is huge. If added transmission were built to and from the Southwest, this source could supply other regions of the country, as well as the growing load in the region.

Utility-Scale Wind

Wind is the world's fastest-growing bulk power electricity source, expanding by 30 percent annually. The technology and mechanics are the same as for small-scale wind—just a whole lot bigger. Utility-scale towers reach some 300 feet or more toward the sky—high enough to capture the strongest prevailing winds. While small wind turbines generate 10 kilowatts or less, the largest turbines today are rated at 6.5 MW, and 10 MW machines are on the drawing boards. These are the offshore turbines—too big to be transported along highways and under overpasses for terrestrial installation.* Turbines such as this are situated off the coasts of Ireland, Spain, and Denmark. In 2007, the United States doesn't yet have any offshore wind farms, but twenty-eight states, representing 78 percent of the country's electricity load, are situated on the ocean front.

Land-based utility-scale turbines range in capacity from 1 to 3.5 MW. In utility wind farms, they're situated so that the efficiency of each turbine is not hampered by the "wake" created by other turbines. Over the years, scientists have learned that air behaves very much like waves. In early wind farms, turbines were placed too close to one another and disrupted air flow.

Another mistake made in early wind farms (namely, at Altamont Pass in California) was placing them in bird flyways. Beautiful birds of prey that can spot a field mouse on the ground from high in the air somehow didn't see the spinning blades and flew into them. This didn't happen often, but it did happen and it happened to some endangered species.

NREL and the National Audubon Society conducted studies to ascertain why the eagle-eyed eagles were flying into spinning blades as they swooped down on their prey. One finding was that birds perched on the towers' open latticework. Changing the tower design to a solid cylinder eliminated the perching hazard. More important, the Audubon Society concluded that if wind farms are

* For perspective on the size, imagine a 6 MW offshore wind turbine. Now superimpose the outline of a 747 aircraft over the rotor. Tip to tip, the wing span of the plane extends over the rotor, reaching the tips of the blades.

not situated in bird flyways, the birds won't be there to fly into spinning blades. Although even one dead eagle is too many, it's statistically accurate to note that more birds are killed by house cats and on vehicle windshields than by wind turbines.

The Good

Utility-scale wind is the only renewable energy technology that's cost competitive with today's conventional bulk power utility technologies. Wind is cheaper than natural gas and new coal (that is, new coal plants either under construction or planned). If externalities were properly accounted for, it would also be cheaper than old coal and nuclear.

In terms of environmental impacts, wind turbines don't require any water at all. Coal and nuclear, on the other hand, require significant volumes of cooling water, and even CSP requires cooling water as well. Water consumption will be an increasingly critical factor in years to come, as stress on limited water resources is expected to become an even more pressing issue than oil.

Wind turbines make a surprisingly small footprint on the ground. They're mounted on steel-reinforced cement pads that, while strong and deep, don't take up much space on the land. Colorado Green reports that the octagonal footing for each of the wind farm's 108 turbines measures forty-seven feet in diameter (and seven feet deep). Less than 2 percent of the project's land area of almost 12,000 acres is used by the actual footprint of the turbine. Farming and ranching continue unimpeded around them (as the many cow pies near the concrete pads attest).

I say farming and ranching, because the very nature of wind farms situates them in rural areas. Utilities and wind developers lease the land from the farmers and ranchers. Individual landowners may earn as much as $3,000 to

$6,000 *per turbine every year.* Fourteen local landowners will benefit from Colorado Green. This supplements—and it may well overshadow—their agricultural income.

In addition, rural counties and towns earn windfall revenues from the property taxes paid by wind developers. Prowers County in southeast Colorado saw its revenues soar by 33 percent in two years when the Colorado Green wind farm came on line. The county had suffered from population out-migration and other economic woes that afflict the U.S. heartland. Then its county commissioners noticed which way the wind was blowing—strong—and attracted a utility-scale wind farm to their county.

Once construction starts, wind farms can be completed within six months. During construction of Colorado Green, local businesses—from the motel to the donut shop to the tamale wagon—boomed. Suppliers—crane operators, cement companies, road graders, all those in the supply chain—also did very well during the construction phase. The economic benefits multiplied when nearby Springfield decided to erect its own turbine, realizing that it, too, could benefit from wind. Springfield cut its own costs by piggybacking and using the cranes, cement trucks, and other construction equipment already in place for Colorado Green, just down the road.

When the dust cleared, the construction hullabaloo died down, and the turbines were up, Prowers County harvested more than $32 million in increased property taxes in two short years. This money supports the local hospital, schools, and other infrastructure investments. In addition, local taxpayers receive a $12 million reduction on their taxes, adding to the attractiveness of communities in Prowers County and perhaps luring new businesses and residents.

Overlay a map of the windiest areas in the Lower 48 on

a map of the counties with the greatest rates of population out-migration (i.e., America's heartland). There is heavy correspondence. This suggests that utility-scale wind can help revitalize failing rural U.S. towns and counties.

Whether local or national, whole new businesses are growing up around utility-scale wind. This includes, for example, windsmiths who climb up 300 feet to maintain turbines and weather forecasters who predict when the wind will blow, so that utilities may more easily integrate this intermittent resource into their systems.

The Bad

The wind doesn't always blow. Utilities assign a capacity value of 30 percent to wind, meaning they anticipate they can use it only about a third of the time. This is probably a conservative estimate, but utilities—with the obligation to serve—plan to meet the need under worst-case scenarios. More troublesome from a utility integration standpoint is the fact that wind is not dispatchable. Coal plants may be only 33 percent efficient, but they're available all the time, generating electricity that can be sent—dispatched—when needed. Currently, there is no way to store wind-generated electricity for times when it's needed.

Some of the United States' windiest areas are not served by the grid, leaving the wind resource "stranded." It will take significant investment to construct the needed transmission lines to carry the electricity from wind farms to markets. In this sense, the construction of added transmission lines is not unlike building farm-to-market roads in the first half of the twentieth century.

The Balance

Transmission and storage will continue to be the pivotal issues limiting development of utility-scale wind. State lawmakers and governors are starting to provide leadership to build the needed wires. In 2004, the Western Governors' Association adopted a goal of building 30,000 MW of "clean and diversified" energy throughout the western region by the year 2030. This goal has a lot to do with the tourism-dependent economies of western states, which require clear air and clean waters. Transmission is recognized as a major constraint, and states are starting to create renewable energy corridors through which transmission can be built before wind turbines are sited—the twenty-first-century utility version of "if you build it, they will come."

Intermittency will be a limiting factor. Key to overcoming this limitation will be the development of technologies that enable safe and affordable storage of wind-generated electricity for dispatch by the utility at a time that meets the system's needs. This will entail both battery and hydrogen fuel storage. In their Wind-to-Hydrogen (W2H2) cooperative research effort, Xcel Energy and NREL are investigating the potential of utilizing wind-generated electricity to split water into its component parts of hydrogen (H_2) and oxygen (O), and storing the hydrogen for later use as a generating fuel.

In the meantime, however, as utilities gain experience with the unique characteristics of the technology, and as the predictive capabilities of wind forecasters continue to improve, utilities will learn that wind is intermittent but predictable. With that information, they can blend this resource into their generating portfolios. Utilities in windy areas may find that there are differing wind regimes within their service territories. It's possible that the wind blows at different and complementary times. Analysis is currently being conducted to assess this possibility. If this spatial diversity turns out to be the case, utilities in some windy areas may be able to increase the capacity factor of wind to more than today's 30 percent assigned value.

In sum, wind will continue be the most affordable utility-scale renewable resource in the short- to mid-term. The amount of wind penetration in any single utility service area, however, will ultimately be limited by the lack of dispatchability—a problem yet to be solved by storage technologies.

Geothermal

Recall our discussion of geothermal heat pumps (GHPs), which tap not into hot heat, but into the earth's constant, moderate warmth about six or eight feet underground. Utility-scale geothermal electricity, on the other hand, is all about heat—sometimes above 100 degrees centigrade. Electricity-suitable geothermal can be hot dry rock, or it can be hot water and steam located in permeable rock. The bottom line is that it must be hot enough to create the steam to drive a generator. That means drilling very deep to access the resource.

The United States' first geothermal electricity plant was built in 1960. Today there are about twenty of them, all located in Western states, where the underground resource is suitable. Most plants are located in Nevada and California, with California accounting for more than 90 percent of installed capacity.

The Good

Geothermal power plants are clean and reliable, the electricity is dispatchable, and they operate at about 95 percent capacity. This is astoundingly high and unmatched by fossil fuel plants and utility-scale renewables.

The Bad

Geothermal is not as "renewable" as solar and wind. This is because underground resources—where the heat of the earth has been trapped for billions of years—are not easily replenished, even if fluid is reinjected. It may take tens or even hundreds of years to replenish this deep underground resource. Having said that, the resource seems to be vast. Identified resources could provide more than 20,000 MW of power in the United States. Resources that are as yet undiscovered could add five times that amount.

The Balance

Geothermal electricity took a back seat to natural gas throughout the 1990s, as that resource was less expensive. Now that the price of natural gas is rising and the federal government has instituted a Production Tax Credit that includes geothermal, more power plants may come on line.

Conclusion

We've come full circle to the quotation that commenced this chapter: There are no silver bullets, but lots of silver buckshot. Technological developments in energy efficiency and renewable energy have given us electricity choices that we didn't have as recently as twenty years ago. Yet, none of them, even energy efficiency, is perfect. Overall, the key limitations of emerging renewable energy technologies are relatively high first costs and lack of availability in the marketplace.

Innovation, development, and maturation of these technologies will continue, perhaps at an accelerated pace, now that we as a nation are waking up to the need to change our energy habits. The cost curves for renewable energy will continue to decline as the cost curves for depleting conventional energy resources continue to rise. Those cost curves will cross within the coming decade, I believe.

So, twenty-first-century technologies, while not perfect, are coming along. But will they penetrate the marketplace? That raises a raft of issues related to our financial and regulatory institutions, perceptions of risk, public policy, attitudes, you name it. These amount to the human factor. Ironically, it's more complicated than technology.

Part II:

Where We Need to Go and How to Get There

What Are We Waiting For?: The Barriers

The most consequential question of the early twenty-first century is who controls the definition of progress.

—Patricia Nelson Limerick

It's been more than thirty years since President Nixon first recognized our military addiction to oil and the national and economic security dangers that it posed in the wake of the 1973 Arab Oil Embargo. It has also been thirty years since Congress created a national laboratory devoted exclusively to researching and developing domestic renewable energy technologies. It was called the Solar Energy Research Institute (SERI). A legislative goal in establishing the laboratory was to create a domestic solar industry. The fact that the mission to achieve this national goal was placed in our national laboratory system, which was created for national defense and weaponry purposes back in World War II days, is telling. Even then, back in the 1970s, we knew: energy is our national security Achilles' heel.

Coal supplies about half our country's electricity; nuclear and natural gas each around 19 percent; old-fashioned hydropower, 6 to 7 percent; and oil, 3 percent. Renewable energy provides about 2 percent or less. The renewable share is disappointing, to say the least. Even worse, DOE's Energy Information Administration (EIA) projects that, while our total energy needs as a nation will increase by 34 percent between now and the year 2030, renewable energy's share will remain roughly the same, proportionately, as today.

I disagree with EIA's forecast. People who understand models say that EIA's models and the underlying critical assumptions don't properly account for the characteristics of renewable energy technologies. Nevertheless, there's no denying that the contribution of renewables to the U.S. energy supply is distressingly small, especially in view of their considerable potential.

If renewables are so great, why haven't they found better traction in the marketplace? In my opinion, the answers—while numerous and complex—boil down to two primary factors: history and marketplace barriers.

A Little History

Prior to the 1970s and the lesson of the oil embargo, a good part of our electricity was generated by oil. The shock to our system of even a short-term inability to secure as much oil as we wanted spurred most utilities to get off oil as a generating fuel. To this day, most utilities that burn oil to generate electricity are in the Northeastern region of the country, where they import it from overseas.

Getting off oil prompted some utilities to invest in nuclear energy. In some cases, this created unprecedented discord with utility customers. Until that time, bulk power had been a relatively new, affordable, and exquisitely convenient service, and customers had been appreciative, complacent, unaware, and quiet. Utilities were the guys in white hats who magically made the lights come on and who provided the air conditioning that made it possible for our hot, humid South to become a major commercial and industrial contributor to the national economy. Consumers neither questioned nor particularly wanted to know how the utilities did what they did. In a very real way, electricity was, and still is, a mystery to those who depend on it for their very way of life.

When utilities took action to reduce their dependence on oil as a generating fuel, nuclear technology—a "peaceful" use of the atom—was emerging as a new and promising base load generation technology. In hindsight, it seems that utilities embarked on the nuclear adventure rather cavalierly (and perhaps prematurely, in some cases). There were several misadventures, some publicized, some not.

In any case, the industry's flirtation with nuclear technology sparked the first significant controversy for U.S. utilities, it created the first discord with their customers, and it created the precedent for consumer awareness and desire for a voice in utility matters. No longer were electric utilities always the guys in white hats. As regulated entities with guaranteed profits, utilities became viewed by some as the largest recipients of corporate welfare in the United States.

Other than getting off oil and trying out nuclear—both of which were significant and occurred throughout the 1970s and into the '80s—not much has changed technologically in the utility sector in the past forty years. In 1967, electric utilities reached their peak of technological efficiency. After that, power plants grew in size but failed to achieve additional economies of scale.

We indulged in a brief flirtation with energy efficiency during the 1980s. This was prompted in large part by the spate of legislation passed by Congress in response to the Arab Oil Embargo, as well as President Carter's stance against energy dependence. But two events dramatically impeded the R & D needed to develop domestically produced renewable and energy efficiency technologies. The same events hindered efforts to inform U.S. consumers about these technologies. The events were (1) the election of

Ronald Reagan as President in 1980, and (2) the clamor in the early 1990s for deregulation of the electric utility sector (which was rooted, to some extent, in dissatisfaction with utility performance as a regulated industry).

Ronald Reagan's Impact on Research and Development

Renewable energy technologies haven't developed as rapidly as one might have predicted when Congress passed the Solar Energy Research and Development Act in 1974. This is a direct function of funding, or, more accurately, lack of it. About three years after SERI opened its doors in Golden, Colorado, Americans elected Ronald Reagan to the presidency. The newly elected president slashed federal funding for the facility in his very first budget and decimated research efforts. SERI's funding was cut by half, and some 500 Institute employees found themselves on the street.

Just a few years earlier, President Carter had declared our energy dependence on hostile overseas energy suppliers to be "the moral equivalent of war." Using energy wisely—through conservation, efficiency, and domestic renewables—was our frontline defense. U.S. oil production had peaked in 1972, and we knew the importance of developing alternatives. The Arab Oil Embargo had caused the unprecedented economic phenomenon of stagflation—a stagnant economy, riddled with inflation—which was the result of both the shortage and the suddenly high price of imported oil. We knew that the threat posed by imported oil was grave.

In response, Congress fashioned a loosely fitted pipeline of technology and marketplace information: new technology information was intended to flow from the national laboratories to DOE and from there to the state energy offices.[*] The energy offices received federal funding to carry out congressionally defined national energy-security goals, and they, in turn, reported the issues they encountered on Main Street. The intent was to inform science and engineering with marketplace feedback.

The election of Ronald Reagan changed the energy conversation, as well as our national priorities. Energy conservation, said President Reagan, was "freezing in the dark." Federal funding that, short years earlier, had been targeted to wean us off imported oil evidently was diverted to another national security objective: outspending the Evil Empire in the arms race.

The Reagan administration zeroed out funding for the newly created U.S. Department of Energy and all state energy offices. Although the states protested the Reagan administration's attempt to dismantle this unique national security network and persuaded their congressional delegations to restore some of the funding, the entire apparatus was decimated. When the smoke cleared from Reagan's first budget submission in 1981, state energy offices had taken as much as a 75 percent hit in one year. Combined with the blow to DOE and SERI budgets, the newborn renewable energy initiative was figuratively smothered in its crib.

No matter what field they work in, researchers will tell you that the pace and success of R & D are a function of funding. On the whole, R & D is a workmanlike, even tedious, endeavor. Success is a function of sustained and

* Actually, Congress created two intermediary institutions as part of the pipeline, both of which no longer exist. Regional Solar Centers fell victim to the first budget cuts of the Reagan administration. The network of DOE regional offices, the sole mission of which was to provide liaison with state energy offices and regional markets for energy efficiency and renewable energy programs, was dissolved under the Bush administration in 2006, even as the president decried our national addiction to oil.

predictable funding. Subsequent budgets restored some funding for renewable energy R & D, but initial funding levels were never reached again, and the overall research effort has been seriously underfunded for thirty years. In addition, the vicissitudes of government budgets and swings in ideological philosophy have made it difficult for scientists and engineers to execute a sustained plan to resolve R & D and manufacturing issues. Another consequence is that, over the years, government-sponsored research has suffered a long-term brain drain, with many good minds giving up and going elsewhere.

An additional impediment has been the pressure imposed by political leaders who sometimes favor, sometimes oppose, the fundamental nature and purpose of the R & D. Some huge successes have been achieved throughout the years, but due to political leanings of politicians who control the purse strings, scientists sometimes have not felt free to talk about them publicly. To this day, the public is unaware of some important renewable energy technology developments and doesn't know about some of their real energy choices.

One would think that private-sector companies, including utilities, would fund energy R & D as a business investment. Although outcomes are uncertain, targeted R & D by those who know the field best and stand to profit from success would seem to be the likely technology path forward in an economic system that purports to reward this kind of informed risk-taking. In fact, however, today's capital markets, while flush, have created disincentives to private sector R & D. Here's why I say this:

1. R & D is expensive, and the outcome, especially the pace at which results can be expected, is uncertain. This makes R & D a risky venture in the eyes of Wall Street. Investors monitor quarterly earnings of the firms in which

they place their money. But R & D is a multiyear investment (multigeneration, in the case of nuclear fusion). Even when R & D yields concrete findings, it then takes years to bring a new product to market. There are no "next quarter" earnings here.

2. After the R & D phase, just bringing a new technology to market is expensive and risky. Most new ventures fail. Only one of a hundred patents is successfully commercialized into products, and four of five new businesses fail within the first five years.[1] A key difficulty is acquiring the financing to gear up production after prototypes have been developed. This is such a hazardous time for new ventures that it has been termed The Valley of Death.

Most new technologies or products are brought to market by small, entrepreneurial firms. Yet, by definition, they lack the financial resources that would give them the staying power to capitalize the venture.

Mention of the deep pockets needed to accommodate this kind of financial investment and risk takes us to the role of electric utilities in underwriting R & D. One would think the utility sector would have a vested interest in conducting this kind of research. In fact, it does. Many utilities pay dues to the industry's research group, the Electric Power Research Institute (EPRI) so that EPRI can conduct research deemed important to the industry. In addition, if allowed by their regulators, individual utilities may be compensated for their R & D efforts. Pacific Gas and Electric (PG&E) in California was a utility R & D leader in the 1980s. However, utility-funded R & D has fallen on hard times in the last twenty years because of another key event, utility deregulation.

The Impact of Utility "Deregulation"

Follow the money.

—*Deep Throat, in* All the President's Men

Whose bright idea was this? Deep Throat's admonition leads us to those who stood to benefit from deregulating the utility sector—that is, merchant generators that had been unable to penetrate the bastions of regulated utility monopoly power and sell their electrical "juice." Natural gas marketers in particular—Enron prime among them—coveted the lucrative electric utility market. Regulated utilities, seemingly bloated and complacent from generations of comfortable regulation and assured profits, turned out to be a surprisingly unformidable target.

Fewer than half of the states actually deregulated their electric utility sector. (Actually, *reregulated* is probably the more accurate term, because regulators continue to provide critical oversight and make decisions that shape the marketplace even in deregulated states.) Nevertheless, throughout the 1990s, the threat of deregulation hung menacingly over utility executives. The consequence was that utility programs were cut if they did not directly contribute to boosting the bottom line and enhancing expected competitiveness in the prospective marketplace. Among the first programs to go were unprofitable R & D efforts.[*]

PG&E was a sad example. PG&E had made its share of costly mistakes. Among them was the utility's failure to staunch dangerous pollution at one of its plants, a tragic

[*] Most, if not all, states created System Benefit Funds as part of the deregulation process. The purpose was to fund efforts not deemed profitable in a competitive marketplace but thought to have social value. Along with low-income ratepayer assistance, R & D is eligible for this kind of funding in many states.

error that sickened and killed local residents and led to the movie *Erin Brockovich.* Nevertheless, PG&E did some smart things in a regulated environment. Chief among them was supporting an outstanding in-house R & D effort.

In the 1980s, PG&E supported a multimillion-dollar in-house R & D program. At that time, Amory Lovins, physicist-turned-guru of energy efficiency, was starting to attract attention to his assertion that the amount of waste in electricity production and consumption was so vast that utilities could offset the need for additional power plants by improving efficiency and minimizing waste.

PG&E's research chief, Carl Weinberg, challenged Lovins to prove his theory that commercially available energy efficiency design and technologies could eliminate 75 percent of electricity consumption at costs comparable to the cost of electricity at the time. The result was a pioneering effort, the Advanced Customer Technology Test for Maximum Energy Efficiency (ACT^2) program.

ACT^2 revealed that for about 7 cents per kWh, energy efficiency could eliminate the need for half the gas and electric consumption in PG&E's service territory. It's important to note that the cap was set by cost and, thus, the 50 percent reduction was an economic limit, not a technological one.

Armed with this information, PG&E announced that, in lieu of building new power plants, it would "mine" efficiencies from its customers. Its regulators permitted the utility to earn a profit off its investments in energy efficiency (rebates and other consumer programs that resulted in hard, verified, sustained savings). Of allowed corporate profits of about $1 billion, $250,000 came from calculated savings through energy efficiency programs during the brief life of ACT^2.

Then deregulation hit in California. The various

impacts of deregulation on vertically integrated utilities are too complicated to enumerate here. The bottom line is that a number of disincentives to energy efficiency were introduced into PG&E's calculus. Moreover, combined with other blows to its business at the time, the utility could no longer afford its first-rate R & D effort. The curtain came down on ACT[2].

PG&E was the poster child for regulated utilities with promising energy efficiency programs at the beginning of the 1990s. In the aggregate, utilities such as PG&E could have accelerated the pace of development and adoption of clean alternative energy technologies in this country. But that ceased with the looming prospect of competitive markets and the concomitant need to focus on the next quarter's earnings.

The key attractions of deregulation were that competition would exert downward pressure on consumer prices and that the breakup of monopoly power would release the floodgates of pent-up technology innovation in a protected industry that had had no incentive to operate in lean and mean fashion.* The fact that no more states deregulated after the initial rush some fifteen years ago suggests that the jury is still out regarding the wisdom of this course. Although some energy providers have branded themselves as "green" and have developed a presence in some markets, technological innovation per se has been slow to hit the markets. In large part, this is due to the sticker shock that, after legislated price caps expire, is hitting electricity consumers. Some markets are seeing prices in the 15 to 20 cents per kWh range.

Consequently, deregulated utilities are no better posi-

* There have been no major technological innovations in coal-fired central station power plants since boiler improvements in the 1950s.

tioned financially to invest in technology innovation than they were in regulated markets. Moreover, due to Wall Street's eye on short-term, next-quarter earnings, utility shareholders are no more willing to absorb the price of R & D and other costs associated with technology innovation than ratepayers are.

Increased bottom-line pressures in a deregulated market may have adverse environmental consequences, as well. Aging, inefficient, dirty power plants are kept in service well beyond their thirty-year life expectancy, in part because they're fully depreciated and help spruce up accounting spreadsheets.

Equally, if not more, worrisome are the potential landmines lurking in reregulated markets. Key among them is a variety of possible threats to reliability. Regulated utilities receive an allowance for routine maintenance and other costs associated with ensuring reliability. In reregulated markets, that's not necessarily the case. Deferred maintenance and lack of infrastructure investment may be unintended consequences of the so-called "free" electricity market.

Investments in energy efficiency are not rewarded in the competitive marketplace, nor is R & D. The statistics regarding private-sector (including utility) investment in R & D are disheartening. Energy has been the least R & D–intensive hi-tech sector of the economy, falling from an investment level of 0.8 percent of sales in 1990 to 0.3 percent in 2004. It's possible that electricity R & D trails the dog food industry.[2]

In addition to Reagan administration policies and utility deregulation, the marketplace itself has erected barriers to entry for renewable energy and energy efficiency. These barriers have combined to impede market penetration of these technologies that would seem to have so much potential for marketplace success and that would serve our twenty-first-century goals so nicely.

Marketplace Barriers

Cost

First and foremost, anything that costs more is going to find hard going in the marketplace. State energy offices and other energy efficiency experts make the case that the higher first costs of, say, a high-efficiency lightbulb, are recouped many times over through savings in energy costs. Nevertheless, added first cost is a hurdle.

In the renewable energy world, utility-scale wind is cost competitive with natural gas and new coal plants today. Other than that, though, first costs of renewable energy are significantly greater than for conventional energy technologies. In 2005, PV costs ranged between 18 and 30 cents per kWh, by comparison to average electricity prices ranging from 5 to 12 cents per kWh.

Property tax valuations also create market barriers for potential renewable energy consumers by adding to the cost. If renewable energy systems, whether solar or wind, are assessed for tax purposes and add significantly to the taxable value of a property, tax liabilities may offset potential savings from reduced utility bills.

Let's say a utility wishes to invest in a wind farm. It might find that regulatory accounting treatments don't make allowances for this kind of investment. Utilities put generating facilities in their rate base, which means that you, the consumer, pay for them. Moreover, although utilities don't profit from their generating fuel costs, they're guaranteed compensation for the fuel costs because they can simply pass them on to you. This is the "fuel adjustment" charge on your utility bill. Consequently, utilities don't have an overwhelming financial incentive to control their fuel costs.

In the case of a wind farm or utility-scale solar, all costs are concentrated in the hardware—the capital costs of the equipment. This can be prohibitively expensive. Yet, the "fuel"—wind or solar—is free, so there's no need to compensate the utility for those costs. As a general proposition, regulatory accounting has not caught up with technologies for which all costs reside in the capital equipment and constitute an up-front financial commitment. Renewable energy technologies are not rewarded for the fact that the fuel is free, even though the costs of conventional fuels are rising.

Addressing the Cost Issues

Public policies to encourage the development of renewable energy are usually aimed at the cost issue. Foremost among them are Renewable Portfolio Standards (RPSs), passed by twenty-three states to date. Although specific RPS requirements vary from state to state, the basic goal and approach are the same all over the country. That is, utilities are required to generate some part of their electricity from renewable energy.

The fundamental goals of an RPS are to (1) jump-start markets for renewable energy technologies and, in doing so, (2) engender economies of scale so that cost is reduced. Those who oppose this policy say it's a mandate and they wish to see a "free market" for energy. However, as discussed previously, there's no free market for energy, and the so-called mature technologies continue to receive billions of dollars in subsidies every year. RPS mandates represent a piddling financial commitment, by comparison.

Government programs, while greatly reduced these days due to lack of funding, also help reduce user costs by providing loans or matching grants. Over the years, DOE has set strategic national goals and competitively awarded grants to public- and private-sector entities to advance the goals.

In 2006, the Bush administration set an R & D goal of cost competitive PV by the year 2015. Reaching this goal will require improved solar cell efficiency, advancements in manufacturing processes, and market growth. DOE funds national laboratories to pursue the R & D piece and has issued competitive solicitations for the remainder, with hefty cost-share requirements. The government part is underfunded, which almost goes without saying. It will be interesting to see if it's sufficient to leverage sustained private sector funding and participation, and if it will result in achieving the program goals.

Similarly, the U.S Department of Agriculture (USDA) awards competitive grants and loan guarantees for rural investments in energy efficiency and renewable energy. Under several titles of the Farm Bill, projects ranging from anaerobic digesters to wind projects are underwritten. USDA recognizes the rural revitalization potential of these technologies and is trying to jump-start rural markets.

EPA also tries to offset cost through grants and assorted facilitating programs and regulatory requirements. This is because the agency recognizes the environmental benefits of clean renewable energy and energy efficiency technologies.

Utility regulators can play a pivotal role in the cost issue. First and foremost, they can demand what a free marketplace doesn't: an accurate accounting of true costs, which is the total price tag associated with a technology, cradle to grave. Mature, established industries such as coal and nuclear benefit from tax incentives, accounting treatments, and subsidies that inure to their benefit. They also benefit from the fact that they're not expected to bear the cost associated with the many impacts of the technologies—the externalities. We, the consumers, pay for these external costs resulting from power plants, but we pay for them in different venues (for example, health insurance and taxes), not through our utility bills.

It should come as no surprise that setting the value of externalities is an extremely contentious issue. Utilities want to set the value as low as possible. In contrast, renewable energy advocates want to set it high, to make renewable energy more cost competitive with conventional fuels. While no one has been able to agree on the correct value for externalities, they can agree that the value isn't zero. Yet, to this day, that's the sole value that has been assigned, by default, all across the United States.

Through their permitting and code functions, municipal governments can change their rules and align incentives to reward minimal fossil-fuel energy use in buildings. The City of Aspen, Colorado, established its Renewable Energy Mitigation Program (REMP) to do this. Aspen combines incentives and punishments ("carrots and sticks") in one of its codes, setting an upper limit on the permitted square footage of new residential construction. If residents want bigger homes, they can have them. This is the United States of America, after all, and it's Aspen in particular. But a fine is assessed for space above Aspen's 5,000-square-foot limit. The money is deposited in Aspen's renewable energy fund and is used to mitigate the impacts of the energy use in the added space. Alternatively, the homeowner may include a renewable energy system in the home (solar water heating or PV) and avoid the fine. By creating a capital pool dedicated to underwriting these technologies, the REMP fund is a way to address the cost issue.

Many electric utilities conduct energy audits for their

customers, or they may encourage their customers to obtain one from a private sector firm. An energy audit is just what it sounds like: an inventory of your home or business to ascertain where you're incurring energy costs and to recommend measures to reduce them.

Part of the audit involves evaluating whether the utility is billing you correctly. This is a desk audit. Residential tariffs and rates are pretty straightforward, but commercial facilities might be billed according to several different tariffs. Rates and tariffs can be complicated, and mistakes are made. Commercial customers may save money just by getting on the right tariffs and being billed on the right rate schedule.

More to the point, however, utility customers can save money by investing in energy efficiency measures. These can range from lighting change-outs to investment in high-efficiency boilers, HVAC, or appliances. Yet, even the best-intentioned consumer might not make these commonsense investments, which would pay for themselves through reduced utility bills, because they can't afford to buy the equipment.

An astonishingly straightforward solution is to let someone else pay. There's a growing commercial enterprise, third-party financing, to do just that. Energy Service Companies (ESCOs) are engineering firms (sometimes with architectural capability as well) that make their profit from utility bill savings resulting from energy efficiency improvements. Using legal instruments called energy-performance contracts, they operate in regulated and reregulated markets alike.

ESCO engineers do an energy audit, calculate the opportunities for energy savings, and purchase the recommended equipment. Obviously, they don't do this out of the goodness of their corporate hearts. The customer signs a contract with them, and the contract guarantees a profit to the ESCO, derived from the stream of energy cost savings.

This mechanism is particularly useful for cash-strapped public-sector agencies, such as school districts and municipalities. It's not as useful for residential customers, as up to now the savings haven't been great enough, nor the revenue stream attractive enough, to entice ESCOs into the market. However, ESCOs in search of a unique market niche could aggregate contracts across a number of homes in one area, bundling the measures and making performance contracting profitable in the residential sector or small institutional or commercial facilities. These days, some ESCOs are starting to provide third-party financing for renewable energy, rooftop solar in particular.

The key reason renewable energy is expensive is the high capital cost of the equipment.* Consumers can help the fledgling renewable energy industry by helping with these costs. Many electric utilities offer voluntary green pricing programs. By choosing to pay a little more on their electricity bills, ratepayers can help capitalize investment in some form of renewable, green energy.

Utilities differ in how they structure their programs, but customarily ratepayers choose how much they want to invest every month and enroll in a program that purchases some amount of green accordingly. One utility, Xcel Energy, sells blocks of wind through its Windsource program. One cool consequence of this program is that if the utility's other generating fuels rise in price, ratepayers on 100 percent wind don't pay the higher fuel adjustment

* PV has special issues, including underdeveloped mass manufacturing capability and the fact that 80 percent of the product is shipped to markets overseas, creating a lack of domestic supply.

charge on their monthly utility bill.* In recent years, when natural gas prices spiked, Windsource customers who were on 100 percent wind received lower monthly bills than their neighbors who weren't enrolled in the green pricing program.

Consumers can also purchase renewable energy credits or certificates (RECs). These are financial instruments that represent the environmental (green) attributes of a renewable energy technology. There are several purveyors of RECs, and the prices vary among vendors and over time. There's a company, Green-E, that certifies the authenticity of RECs and the veracity of the attributes they purport to represent. Nevertheless, you need to be an informed shopper if you do this on your own.

Increasingly, however, utilities are offering REC programs under which you may elect to make purchases. The utility does the shopping, and it charges an administrative fee to handle the transaction. This is a convenient way for individuals to help capitalize the development of the renewable energy industry and, accordingly, help reduce consumer costs, even if their utility doesn't have its own renewable energy program. (Remember, not all areas have a suitable renewable energy resource base.) This helps defray the early costs of development and helps jump-start the industry.

The downside is that, if you have renewable energy resources in your area, RECs may create the unintended consequence of impeding actual renewable energy development. It's cheaper and easier to purchase RECs than, say, to erect wind turbines.

* With this type of program, you don't know if you're getting electrons generated by wind or natural gas or coal. But you feel good because, through the premium you voluntarily pay, you contribute to the worthy cause of developing additional renewable energy.

Artificial Barriers Resulting from Regulatory and Wall Street Disincentives

Anyone who tells you there's a free market for energy is either fibbing or has been duped by the self-serving babble of the entrenched energy industries that avoid competition by barring others from market entry, even as they continue to profit from their own array of subsidies and accounting treatments.

By one estimate, the world's coal industry receives $53 billion annually in subsidies, even though this is a mature industry that should be able to compete in the marketplace under its own steam, as it were. Natural gas receives $46 billion worldwide in subsidies, the electricity industry gets $48 billion, and nuclear is handed $16 billion. In contrast, a mere $9 billion is the price tag worldwide for subsidies to the renewable energy industry.[3]

Ironically, although PV and wind technologies were developed largely in the United States, we lag behind the world's late-developing countries in adopting them. From New Guinea to China, technology consumers are leapfrogging twentieth-century electricity technologies and going straight to distributed wind, solar, and biomass gasification.* This is because, on the whole, late-developing countries haven't built the infrastructure of central station power plants and transmission grids that we have in the United States. Because our utilities purchased this capital equipment, it's a sunk investment and seemingly their cheapest operating alternative (especially since they don't pay for externalities). This has made utilities reluctant to

* China, blessed with abundant coal resources and with an immediate need to power its burgeoning economy, is investing heavily in coal. Both China and India are expected to build nuclear power plants, as well. Nevertheless, China has made significant commitments to efficiency and renewable energy.

add other kinds of equipment to their systems.

Through the years, our utilities developed the expertise to run systems that consisted of big power plants and big wires. Utility decision makers, mindful of the obligation to serve, are hesitant to adopt new technologies and new operating models. They're unfamiliar with the technologies, don't know how to blend them with their systems, and don't have employees trained in the unique issues posed by renewables (for example, intermittency and the need for utility-supplied backup power, which, by the way, is not a 1:1 ratio of renewable to conventional, a claim that some utilities continue to make).

Regulators are complicit in favoring yesteryear's technologies. Regulated utilities are allowed to earn a stipulated profit margin, and the regulatory system defines profitable activities. Regulators are likely to calculate profit margins based on utilities' investment in "plant and equipment." Utilities that invest in rebates for consumer investments in energy efficient lightbulbs might be permitted to recoup the costs of the program but not earn much, if any, profit. This financial incentive might not be lucrative enough for utilities to make the investment.

Wall Street also creates barriers to market entry for renewable energy. Bond houses rate electric utilities, scoring them in terms of riskiness as investments. This affects the cost of money that utilities can attract: the lower the bond rating, the higher the cost of money. Consequently, utilities behave corporately in ways designed to please the bond rating houses. Under the current system, investment in capital equipment—that is, power plants—is valued by bond rating houses. A utility's desire to improve its rating may drive it to build additional capacity, which might not be warranted, strictly on the basis of load.

This creates a financial burden for consumers, who must

pay for the new plant in their utility bills. It also creates a disincentive for the utility to invest in energy efficiency or cost-effective renewables, because it has to sell electricity from the new power plant in order to recoup that investment. Under most rules, utilities are compensated for, and profit from, selling *electrons* rather than *electricity services.*

A powerful but sometimes overlooked marketplace disincentive to rooftop solar technologies lies in restrictive building codes and homeowner association covenants. Local permitting processes can present formidable barriers for individual would-be electricity generators who don't have a team of lawyers and engineers at their disposal. For example, a restaurant owner, inspired by a painting of the Moulin Rouge that featured a windmill, decided to put two forty-foot-tall wind turbines on the roof of her downtown Denver café. She learned that the City's zoning department had no guidelines for this kind of installation. Before she could obtain the necessary permits, the City had to create them. This involved working with City staff and making an appearance before city council—all of which involved time and effort—but the restaurant owner persevered and got her wind turbines. This story had a happy ending, but it's because the City wanted it to. Often, these stories don't have happy endings.

Arizona, a state obviously blessed with an abundance of solar energy and home to numerous wealthy retirees, has been slow to embrace rooftop solar technologies because homeowner association covenants prohibit structures on roofs. Many retirees could be expected to afford the purchase of a solar system, and a solar system would make tons of sense in the desert, but artificially imposed policies have slowed the development of this market. Under DOE's (now defunct) Million Solar Roofs program, this issue was addressed in several communities.

Addressing the Regulatory and Wall Street Issues

For the viability of electricity investment decisions, Wall Street needs to learn more about clean twenty-first-century energy technologies and how they can be employed reliably and affordably as part of a utility's generation mix. As of now, it's apparent that decision makers in the financial communities don't know much about the performance characteristics, cost, and risks (or lack thereof) associated with renewable energy and energy efficiency.

Similarly, bond rating houses and other players in the investment community need to reevaluate how they assess risk. What are the real risks associated with electricity generation and transmission? Are they changing over time? What are the most significant ones? Depending on the answers to these questions, financiers might conclude that their current investment and bond rating criteria don't induce optimal investments, either from a risk perspective or in terms of the common good. Some people in Colorado, for example, argue that a coal plant currently under construction is not needed but is being constructed mostly to enhance one utility's bond rating. Regardless of the merits of this or any other argument, the point is that *Wall Street can change the rules to incentivize investment behaviors* that are generally aligned with our country's national well-being.

Part of the calculus, no doubt, will be how Wall Street assesses the likelihood that Congress or state legislatures will impose a carbon tax or similar financial policy aimed at reducing carbon dioxide emissions. If the financial community weighs the likelihood of such government action as probable, utilities may have a hard time attracting financing on favorable terms for new coal-fired power plants.

Perhaps, as it assesses risk, the financial community will also start to demand generating-fuel diversification for risk mitigation, just as it recommends diversifying our stock portfolios.

Regulators have incredible power, which they can use for good or ill. If they alter their rules to provide meaningful incentives, they can change the entire electricity generation model in this country. If continuing to build large central station power plants (nuclear or fossil fuel) is the desired behavior, regulators can continue to allow utilities to earn a profit on plants and equipment and can otherwise structure the financial incentives to reward this kind of utility investment behavior.

If, on the other hand, regulators believe that utilities could reduce or avoid the need to build costly power plants by investing in energy efficiency and renewables, regulators can provide the needed incentives. If regulators and other decision makers determine that distributed generation enhances our homeland security and hedges risk, they can change the incentives to align with public policy and the common good. Regulators can change their rules.

Regulators can allow utilities to earn an attractive profit on investments in energy efficiency and renewable energy, and they can require utilities to create time-of-day pricing programs, rewarding customers with cheaper electricity at off-peak times and charging them higher rates for peak-time purchases. Following straight market principles, one would expect customers to shift their loads to the extent they are able.

This would nudge the utility's load shape toward its horizontal straight line ideal by reducing the peaks and filling in the valleys. Inasmuch as utilities build additional power plants to meet peak demand, shifting loads to clip the peak—i.e., load management—can reduce the need to build additional plants.

Regulators can decouple utility profits from the sale

of kilowatt-hours. *Twenty-first century utilities should be in the business of providing energy services, not merely peddling electrons.* This is not as simple and straightforward as it might sound. Regulated utilities, with their protected profit margins and monopoly status, must avoid anticompetitive practices.

Nevertheless, they can fund rebates for purchases that will benefit the entire system. High-efficiency appliances benefit not only the individuals who take the rebate and make the purchase. They also benefit the merchant who sells the appliance and the person who installs it. Critically important, but not well understood in the twentieth-century utility industry, is that rebates benefit *all* utility customers, because—in the aggregate—investments in energy efficiency reduce the need for power. Measures that reduce overall electricity consumption or clip peak demand can defer or avoid the need to invest in expensive new power plants, purchased power, or high-priced spot market purchases.

The regulatory argument that is invoked by twentieth-century utility spokesmen and regulators to oppose such utility rebates is that, by their very nature, energy efficiency products reside in privately owned dwellings and they benefit individual residents or businesses. Consequently, helping purchase those products through utility-funded programs such as rebates amounts to what is called a cross-class subsidy (that is, one group of customers pays for another's benefits) or underwriting "social" goals. This is presented as an equity issue. Moreover, if some recipients of program services would have made the purchase even without a utility program, then they're getting a free ride and it's not fair to the other ratepayers. In utility-speak, this is the "free rider" issue.

What these arguments fail to account for, however, is

that utilities build new power plants in order to serve load growth. Growth that is significant enough to warrant a new power plant is engendered by population in-migration and commercial and industrial growth. If one applies the aforementioned logic regarding rebates to new power plant construction, only newcomers on the system should be billed for the costs of new plants and equipment. Arguably, they're the only ones benefiting from the added investment. Those who were already on the system were doing fine and didn't need the additional plant. Therefore, one could say that the newcomers are being subsidized.

Xcel Energy is a case in point regarding the power of regulators. It has made notable investments in energy efficiency in its Minnesota service territory. Moreover, a spokesperson for Xcel estimates that its energy efficiency programs have saved it (and Minnesota ratepayers) the cost of eight or ten 250 MW power plants. Xcel reports that it costs less than 2 or 3 cents per kWh to do this—a cost far less than any new power plant. Xcel operates in a number of states throughout the Midwest, yet only in Minnesota has it made this notable level of investment in energy efficiency. Same company, different rules and regulators.

We need the massive amounts of electricity that yesterday's power plants continue to generate. However, by providing the right incentives, we might be able to defer or, in some service areas, perhaps even avoid altogether the cost of new power plants.

In terms of building codes and covenants, jurisdictions that award permits and other legal permissions related to construction can agree to fast-track energy efficiency or customer-sited renewable energy projects. This is happening in some California towns.

Importantly, homeowner associations can change their covenants to permit solar. Instead of dispatching the paint police to strong-arm their neighbors and achieve cosmetic conformity throughout the community, HOAs could use their (truly awesome) powers to encourage commitment to the genuine well-being of the community, both today and for the future. Even the prettiest neighborhoods will fall into disrepair if the inhabitants are forced to divert increasing proportions of their household budgets for unnecessarily high utility bills. HOAs could be a positive force by encouraging energy efficiency and on-site distributed renewable energy. Today, that would be rooftop solar. Who knows what possibilities tomorrow will bring.

The reason I'm directing some ire at HOAs is because they represent the *attitude* barrier, and this may be among the most difficult to overcome. Many HOAs are situated in sunny locales and are populated by relatively wealthy inhabitants who could afford to purchase PV and solar water heating systems. The only reason these communities aren't burgeoning markets and solar showcases is because someone decided solar isn't *pretty*. Some people think CFLs are ugly, too. We must change our notion of beauty. To the extent that yesteryear's attitudes inhibit us from making investments that are in our abundant self-interest (for reasons of cost control, future risk hedging, environmental, you name it), *yesterday's notions of "pretty" are today's luxury and tomorrow's burden.* Do not compare a wind turbine to a tree. The intellectually honest comparison is to a coal train or the cooling towers of a nuclear power plant. Which is "prettier"?

The Utility Industry's Own Artificially Imposed Barriers

It takes guts and a pioneer's independence to go "off grid," and few people actually do it. Going off grid means that you meet your own electrical needs, and because you eschew a utility wire running to your house, you don't have the option of turning to the utility for backup power if you need it.

The more common route for independence-minded electricity users is to meet *some* of their electricity needs, relying on the grid as supplier for the rest. If they construct net- or near-zero energy buildings, these consumers might be able to supply all of their needs, on average, over the course of a year. But they remain connected to the grid for those times when they can't generate sufficient electricity on their own. In these cases, the grid provides backup power, and the utility functions as an electricity insurance carrier.

Most people choose to connect to the grid because battery backup can double the cost of the system. Inasmuch as regulated utilities have the obligation to serve, they must provide the backup or supplemental power you seek. But they don't like it, and understandably so. The utility must generate enough electricity to send to you in case you need it, yet it can't plan on how much you'll need and, thus, how much it can sell.

Consequently, utilities may erect barriers for anyone seeking this backup service. Until the recent advent of small, on-site distributed systems such as rooftop PV, the only customers that wanted this service were large industries. They were able to generate some amount of their own electricity through cogeneration, a process through which waste heat is recaptured from industrial processes and used to generate electricity, or vice versa.

Large industries have been capable of cogeneration since the beginning of the twentieth century. However, it was only in the aftermath of the 1973 Arab Oil Embargo that public policy caught up with technology and Congress passed a law requiring utilities to permit companies to interconnect and sell cogenerated electricity back to the utilities. This was the Public Utility Regulatory Policies Act, better known by its unfortunate acronym, PURPA.

Arguably, the provisions of PURPA conflicted with the established regulatory philosophy of rewarding utility investment in plant and equipment. If utilities were forced by law to purchase large amounts of power from on-site generators and cogenerators, they might be forced to scale back their forecasts of future capital needs and the associated revenue permitted by regulators. This could pose a risk to their financial health.

Consequently, utilities did two things. First, they tried to remove the financial incentive for large customers to cogenerate. Each utility set an avoided cost rate—that is, what it would have cost the utility to generate the electricity offset by the cogenerator. Because utilities supposedly achieve economies of scale and because they want to pay cogenerators as little as possible, avoided costs are as low as utilities can persuade their regulators to sanction. This is what cogenerators, and small on-site generators, are likely to receive for electricity they feed to the grid.

Second, utilities developed interconnection standards, ostensibly to protect the integrity of the transmission system. Interconnection standards are exactly what they sound like—the rules you must follow in order to connect your system to the utility's. Among the requirements, for example, might be expensive liability insurance.

In fairness, it must be said that the transmission system is, at its core, a quirky highway of moving electrons that must be kept in exquisite balance at all times, or else the system, quite simply, crashes. In addition, there are serious risks in keeping wires "live" by transmitting electrons through them. There have been instances, for example, of individuals firing up diesel generators when power lines were downed by storms. Utility linemen, unaware that the lines were live, were electrocuted as they worked to restore power. So we can understand the grave risk of letting everyone and their dog mess with the wires.

Nevertheless, in trying to manage the interconnection risk, utilities impose a costly paperwork burden on would-be generators. Industrial cogenerators have long complained about it, even though they tend to be companies with significant resources and legal teams to match the utilities'. Understandably, this kind of burden is a deal-killer for small power producers, such as residential generators. Small power producers lack the legal and accounting staffs to file the voluminous certifications and other paperwork often imposed by the utilities. In this way, interconnection standards become a formidable disincentive to distributed generation.

If you make the investment in an on-site generating system and if you endure the hassle of comporting with your utility's interconnection requirements, you probably would like to be compensated, at least in part. Your utility might or might not be willing to do so. If so, it would be through its net metering policy. The meter that ticks forward as you receive electricity, recording your usage for billing purposes, can also run backward if you feed electricity back into the line. If your utility is willing to purchase the electricity from you, you enter into a net metering agreement.

In its true form, net metering uses one meter—yours—and permits you to be paid by your utility for excess electricity you feed back to the grid. In residential situations, you can expect to use all or almost all of the electricity you generate. This is because a residential roof is not large enough or strong enough to accommodate enough solar panels to generate large amounts of electricity. Nevertheless, there may be sunny times when you generate more than you use, and your utility should be willing to compensate you for that excess amount.* Hence, the term *net.*

Utilities differ in how they calculate the net. A common approach is to settle up at the end of the year. The utility might pay you for the net amount it purchased from you (although it probably won't pay at the retail rate at which it *charges* you); more likely, it will give you a credit toward your future purchases from the utility. One rural electric cooperative in Washington, the Chelan public utility district, combines net metering with a voluntary green pricing program. Chelan creates an annual pot of cash, capitalized voluntarily by members who wish to contribute to green power in their service area. At the end of the year, Chelan divvies up the pool and distributes it on a pro rata basis to the net metered green power generators in the district.

Over time, the electric utility industry has grown to the point where it's the largest industry in the United States. The sheer size of the industry, its status as a protected (i.e., regulated) industry, and the technological complexity that makes it opaque to most consumers have combined to vest the utility industry with quasi-governmental powers. Utilities can use these powers to impede market entry by upstart distributed-energy technologies. A few utilities are actually quasi-governmental in the sense that they have government backing as institutions and receive government funding. The big dogs in this category are Bonneville Power and the Tennessee Valley Authority.

In addition, there are numerous smaller entities, including regional river authorities and rural electric

* Which, I might add, it can sell at retail rates, probably having purchased it from you at the lower avoided cost rate.

cooperatives. Also called REAs (Rural Electric Associations), rural electric cooperatives have emerged as a bastion of yesteryear utility thinking. They were established by federal law in the 1930s in order to electrify rural areas.* REAs are, in philosophical underpinnings and in theory, the epitome of grassroots democracy. Co-op members aren't customers or ratepayers; they're member-owners. They hire professional staff to run the utility, and they elect a board of directors from among their neighbors to oversee its operations. Most co-ops have fewer customers than IOUs, and they usually purchase electricity rather than build and operate power plants on their own.

Importantly, the ideal of co-ops as grassroots democracies means that sometimes they have little or no regulatory oversight. Even in states with regulated utility sectors, REAs might not fall under regulatory jurisdiction of any kind.

This is a generalization, but I think it's fair to say that co-op members are no more interested in the inner workings of their utility than ratepayers in investor-owned utilities. Without term limits, those who are willing to serve on co-op boards of directors tend to become ingrown amongst themselves, and they form cozy relationships with the co-op's wholesale providers. In rural America, these wholesale providers overwhelmingly generate with coal, some of them even owning their own coal resources.

The result is sadly ironic. The U.S. heartland could revitalize failing rural economies and could stem population out-migration with renewable energy "cash crops." Farmers and ranchers in windy areas with transmission access can

* Private-sector electric companies were unwilling to extend their wires to sparsely populated rural communities, because there simply were not enough customers to make the investment profitable. Even today, REAs average seven customers per mile of wire, compared to around thirty-five for IOUs.

lease their land to utilities and wind developers for a hefty price, as much as $3,000 to $6,000 per turbine *every year.* They can continue to farm or run livestock around the base of the turbines, so this is a twofer beyond the wildest imagination. In addition, utilities or energy providers pay generous property and other local taxes, infusing the public coffers of local economies. In addition to wind, biomass and solar opportunities also abound in rural areas.

Yet, the local co-ops, captives of long-standing relationships and contracts with coal-based wholesalers, resist the addition of renewable energy to their portfolios. Sadly, local board members may be at odds with their neighbors. Downright nuts, though, is the fact that board members themselves are often farmers and ranchers, and they're failing to act in their own best interest.

In fairness, we should acknowledge that long-term contracts, which are what coal companies have negotiated with REAs, are not simple to modify. On the other hand, however, awful sticker shock awaits rural co-op members (as well as IOU and municipal utility ratepayers) when those contracts expire. Consumers whose electricity is generated by coal have been getting something of a free ride because their utilities locked in lower prices some time ago. Spot market prices are a good indicator of future contract prices, and those are doubling and tripling these days. Co-op members aren't well served if they don't start to diversify their electricity portfolios and reduce their financial exposure.

Not trivial is what Hermann Scheer refers to as the "cultural hegemony of the traditional energy business," or, in more sporting terms, the utility "home team advantage."[4] The bottom line is that the consuming public, grateful for the ease with which we obtain our energy supplies (whether electricity, heating, or transportation fuels) and not wanting to be bothered with the details, favors the devil we know over the one we don't.

There are memorable examples of big mistakes made by the utility industry: massive cost overruns or cancellation of nuclear power plants and blackouts of entire regions are two examples. But on the whole, day in and day out, the system works very well, and we interpret that to mean it ain't broke. And if it ain't broke, don't fix it.

Addressing the Utility Industry Issues

A new model, supplementing power plants with distributed energy generation and renewable energy technologies, is the devil we don't know. If the utility industry tells us it's okay, we will believe. Utilities are the experts and we don't want to become experts ourselves. To date, it has been against the vested interests of the utility industry to support this kind of paradigm shift. If utilities change their corporate minds on this subject, the consuming public will follow. Regulators can be instrumental by changing the rules the utilities must follow.

The willingness of the consuming public to rely on the leadership of assumed experts makes the utility industry and its regulators what we call influencer groups. If electricity providers and regulators are persuaded that a paradigm shift would in fact be a good thing, needed changes will be made. Inasmuch as most states continue to regulate their utilities, this is a clarion call for utility regulators to create meaningful incentives to nudge regulated utilities toward making needed changes. That is, regulators should change the rules.

Barriers Resulting from Inaccurate Price Signals

Whether we're talking barriers or incentives, it almost always comes down to one thing: money. Individuals or commercial enterprises may make investments in energy efficiency (for example, changing out lightbulbs) if their electricity costs are going up. They have a clear financial incentive to try to manage those costs.

But what if the person who pays the bills isn't the same one who makes decisions about energy-related purchases? This is called the principal-agent, or split incentives, dilemma. Rental property is the prime example. The owner doesn't customarily work or reside in the property and, consequently, doesn't pay the electricity bills. He or she is, in economic terms, indifferent to the costs of heating, cooling, and powering the structure. Unless it's a renters' market and the landlord is looking for a unique marketing niche, the landlord is unlikely to invest the money to make improvements, because he or she doesn't feel the pain of the utility bills.

In contrast, the inhabitant, whether a commercial enterprise or an individual renter, is keenly aware of the costs of operating the structure, because he or she pays the bills. If space conditioning and electricity costs are rising, the inhabitant would be inclined to make improvements in the structure—perhaps added insulation or high-performance windows. But the inhabitant doesn't own the structure and is therefore unlikely to make the needed investment to improve its energy performance.

Addressing the Price Signal Issues

Again, it's a case of getting the incentives right and making sure they elicit behaviors that are desired. In the case of split incentives, meaningful rewards (preferable to punishments and penalties) can be instituted to persuade all parties to do the "right" thing. Perhaps landlords could be given tax incentives to make energy efficiency improvements and inhabitants could be given some kind of rebate or other incentive. New York City has announced plans to work with the financial community to upgrade the city's inventory of existing building stock (950,000 buildings, most of them privately owned) and improve energy efficiency with capital improvements such as high-efficiency furnaces and air conditioning, insulation, and improved doors and windows. Because of the principle-agent dilemma, New York will need to create meaningful incentives for landlords, or it may create some regulatory requirements—for example, creating energy performance standards for rehabilitation projects or when property changes ownership (typically ten years for commercial space, somewhat less for private homes).

Easily accessible and smartly targeted information should be made available to all decision makers, whether individuals or homeowner associations. A good way to do this is through their own communication channels (newsletters, websites, and so forth). But the question of *who* should do this remains. It seems to me that providing objective, accurate, timely information is a low-cost and appropriate role for government.

Market Barriers Resulting from Insufficient Information

Classical economists assume perfect information in the marketplace. I'm not sure that has ever been the case in the real world. But at today's rapid pace of technology development, that is definitely not the case.

Ask production home builders why they don't incorporate energy efficiency into their dwellings, despite the fact that energy efficiency measures are commercially available, pay for themselves in energy savings, and can be relatively painless when rolled into a mortgage. The builders respond that the consumer doesn't demand it. This is as frustrating a chicken-and-egg issue as you're likely to find. Consumers don't demand energy efficiency because they are unaware of the financial benefits and they don't know, specifically, what they should be demanding. They're unaware because they haven't been informed. In our hi-tech society, in which technological knowledge tends to be narrow and reside with the experts, consumers don't even know what questions to ask about many products. They assume that the experts—builders, in this case—will inform and guide them. Unfortunately, in terms of energy efficient design and construction, as well as on-site renewable energy, it's just not happening to any significant extent.

Another reason builders don't incorporate energy efficiency, passive solar design, and rooftop solar technologies into their homes is that architects and builders themselves don't know much about these options. Woe (and kudos) to the individual who tries to construct a building that is, cradle to grave, sustainable. He or she will have a hard time finding a knowledgeable architect and an equally hard time obtaining supplies and products. As Kermit the Frog said, "It's not easy being green."

DOE, state energy offices, and some electric utilities have tried to provide needed information. As government budgets shrink, however, there's less funding for the so-called soft programs such as public information and outreach. More and more, government officials are demanding metrics that track measurable outcomes of government programs. Public education and outreach don't produce measurable results and are cut back accordingly.

Addressing the Issues Related to Insufficient Information

This gets us back to the notion of influencer groups and their impact on consumers. In terms of electricity, influencers are utilities, other power providers, and regulators. In the building sector, builders and developers are foremost among the influencers. Consumers don't demand energy efficiency or distributed renewable energy because they don't know enough to do so. Builder-developers can influence buyer decisions by offering and marketing energy efficiency and renewable energy features, just as they do granite countertops.

Remember the Scripps Highlands story? Even though the homeowners with PV and solar water heating came to love those features once they saw their utility bills, the fact of the matter is that some of the homeowners didn't know they owned these systems until it was pointed out to them. These were the unwitting adopters. They were unaware because the people selling the properties didn't tell them, and a teachable moment was lost. The builder-developer, while undertaking a heroic pioneering effort, didn't seize the opportunity to educate and didn't use his bully pulpit as an influencer. In addition, electricity providers can influence builder-developers to incorporate energy efficiency and distributed renewable energy just as, in the past, they provided incentives for developers to build all-electric homes.

The most critically important influencers are the money people—bankers, mortgage lenders, and financiers of all stripes. If they're confident of a technology or a product, they can make capital available to underwrite it. ESCOs could not exist but for the banking community's understanding of the suite of technologies ESCOs employ and the lenders' confidence that the technologies will repay the investment through energy cost savings.

Similarly, mortgage lenders need to understand the cost savings that result from energy efficient homes. Energy efficiency can lower monthly operating costs significantly—so much so that embedded measures, whether passive solar design, optimal insulation levels, high-performance windows, or whatever, should be considered when lenders qualify potential buyers. A legal instrument exists today to capture this relevant financial: an energy efficient mortgage. However, many lenders seem unaware of it, home buyers have never heard of it, and the instrument itself might not fully reflect the contribution of energy efficiency to monthly cash flow. The financial community can make sure its front-line representatives who deal with the consuming public are aware of the benefits of energy efficiency and on-site solar, as well as the financial instruments that can help make these options affordable. They can use their power as an influencer at critical decision points, such as negotiating a mortgage.

Other critically important influencer groups, as far as the residential sector is concerned, are real estate agents and appraisers. How is it possible that some homeowners in Scripps Highlands didn't know they had PV on their roof? One can only speculate as to whether the sales staff, real estate agents, and even the appraisers understood the potential value of the structure on the roof, or if they were even aware of its presence. Conversely, and distressingly, it's possible that they were aware of it and assigned it a negative value as an eyesore. In years past, many an owner of a rooftop solar water heating system saw his or her property devalued because it wasn't considered pretty.

Government—whether federal, state, or local—can provide easily understood, easily accessible, timely information geared toward specific influencer groups and decision makers. Government isn't a vendor and doesn't have a vested interest in peddling one product or technology. Government can be objective, and can provide information in venues and through channels that it knows are used by its citizens.

Also, government can provide objective information

about specific products and technologies. It can label the performance of electricity-using equipment. Government energy ratings started with refrigerators and other household appliances. Now there are Energy Star ratings for office equipment and buildings as well.

Importantly, government can lead by example. The City of Chicago grows plants on its downtown office buildings—living, green roofs. This helps keep the buildings cool (reducing air conditioning costs) and offsets some amount of CO_2 emissions. Chicago also installed PV on its office roofs and arranged for PV installations on schools throughout the metro area.

Like government, the private sector can provide useful information in ways not tied to selling a product. A new industry of home-energy raters has grown up. These are independent firms that inspect individual homes and evaluate their performance. Importantly, energy raters ascertain if homes perform at the levels promised by the builder.

Some electric utilities use the "communication" venue of their monthly bills to disclose their generating fuel mix. This enables customers to understand where their electricity comes from and what the impacts might be. It provides electricity users an opportunity to think for themselves about the externalities associated with their electricity use.

Finally, and importantly, individual consumers can take responsibility for their own behavior and purchase decisions. In these days of the Internet and powerful, easy-to-use search engines, most information is reasonably accessible. Consumers just need to remember the age-old warning, caveat emptor—let the buyer beware. Do your homework and tap into objective and knowledgeable information sources—perhaps government or nonprofit associations that don't have the conflict of interest associated with trying to sell you a product.

Industry Fragmentation and Lack of Capacity

The fact of the matter is that there is no true "energy-efficiency industry." It's highly fragmented, ranging from the occasional architect who designs energy efficient structures to the retailer who sells CFLs. There are the appliance manufacturers, some of whom offer efficient versions of their standard product line. There are the retailers. There are the rare green builders. There are the insulation manufacturers and retailers. There are the window people.

They don't speak with one voice. They aren't a cohesive industry, and, consequently, they lack political clout. When they exhort you to buy their products, they sound like vendors hawking their wares. They don't sound like they're informing the consuming public, even if they actually are, because they're trying to sell a product. This is the fragmentation issue.

Then there's the labor capacity problem.

I remember leaning against a window in a newly constructed facility at NREL. The structure was intended to be a living laboratory of energy efficiency and passive solar design. Perched on that windowsill, I felt a decided winter chill. It was, literally, butt cold. When I complained to the researchers who had assisted in the structure's design and had overseen construction, they explained that the contractor had failed to install the thermal break called for in the blueprints. This had happened on a day when the researchers had to be elsewhere and had not been on-site to monitor installation of the windows. The contractor evidently didn't have sufficient knowledge or appreciation of the issue, or this wouldn't have slipped through the cracks, as it were. This illustrates the labor capacity issue.

If rising electricity rates and other factors suddenly create a large and sustained market for energy efficiency and renewable energy, who will meet the need, and how will they do it? Installers need to be properly trained, and if you live in rural areas, the chances of finding a solar retailer are currently slim to none. Once more, the capacity issue.

Addressing the Fragmentation and Capacity Issues

The U.S. Department of Labor and community colleges across the country are starting to take a serious look at the U.S. workforce and question whether it's trained to meet the changing needs created by twenty-first-century technologies of all kinds. In some cases, the industries themselves are getting into the act, as they struggle to establish professional credibility and, in the case of the solar water heating industry, reclaim it. The poorly structured solar tax credits of the late 1970s attracted an unseemly number of fly-by-nights to the solar water heating business. Many companies were as short-lived as the tax credits. Nevertheless, unskilled installers and irresponsible firms dealt a lasting blow to the solar water heating industry's professional reputation and the credibility of the technology. This was a painful lesson in the importance of qualified installers.

The labor side of the issue has been addressed by a number of states. However, as one would expect, each state handles installer certification and training differently. As the PV industry matures, the industry itself seeks standardized installer training and certification requirements. The North American Board of Certified Energy Practitioners now tests for mastery of PV and solar water heating installation, and is adding small wind to its certification program.

No doubt the labor capacity issue is related to the youthful age of the renewable energy industries. The fragmentation issue related to energy efficiency products and services is something else. It will be interesting to see if someone can wrestle this one to the ground.

Concluding Thoughts

The preceding list of barriers (and efforts to overcome them) is by no means exhaustive. The purpose is merely to introduce you to the issues that, taken together, help explain why these technologies aren't more prevalent in the marketplace and why you as a consumer haven't heard more about them. Now, however, I anticipate—and see evidence of it—that strategies to address the issues are starting to have impact. Market conditions and drivers are changing as well. Key among them are the rising costs and growing consumer awareness of the issues associated with twentieth-century fuels and technologies.

In candor, any number of public policies intended to do good things have not always been effectively structured. Quite simply, there is a learning curve on the policy side to parallel the technology learning curves.

The Future: It's Already Here

We cannot know with absolute certainty, so we do nothing … The essential human dilemma is that all our experience is in the past and yet all our decisions relate to the future.

—Richard D. Lamm

Should you find yourself in a chronically leaking boat, energy devoted to changing vessels is likely to be more productive than energy devoted to patching leaks.

—Warren Buffet

These two statements illustrate the yin and yang of planning for our electricity future. It's painful to contemplate major, technologically disruptive changes, because we don't have experience with the new and different systems that technology allows. We also have no precedent for the untried solutions that tomorrow's problems demand and that we must address today. All in all, we would prefer not to act, especially not to *change.*

At the same time, we hear from an esteemed American (and one of the most successful businessmen our capitalist system has produced) that, when things start to break, it's time to take decisive action. Action should be taken even if it's as disruptive as "changing vessels" rather than fixing what we have and are familiar with.

Believe it or not, despite the formidable barriers that must be overcome in order to ensure a reliable and affordable electricity future, it may be that our future is so bright that we have to wear shades. It all depends on how we approach the institutional and financial barriers. The optimistic view, which I personally prefer, is that barriers are merely problems in search of solution.

Pathways toward solutions are starting to emerge. Like any new path, they must be walked before the best routes become clear. Importantly, we'll need inventive genius of all kinds, and this is the United States' strong suit.

Technology innovation will not be enough. Big changes will need to be made, purposefully and proactively, in the entire institutional and financial apparatus—a massive undertaking. And in order for that to happen, our values, our mindsets, our notions of beauty, our perception of risk—all must change. Can we muster the political courage and national will to ease into the revolution? That is the $64,000 question.

There's no doubt that we possess the technology tools to nurse our electricity generation and transmission system into the twenty-first century. We need to do this, and we need to start now. We can patch our leaky twentieth-century energy boat and continue to sail it for a while. In fact, unlike Buffet, we can't afford to buy new vessels whenever we want. But we need to start shopping and thinking through our options, because we're starting to take on water. Purchasing a model exactly the same as the old boat would be akin to hanging curtains in the wheelhouse as the vessel sinks beneath the waves.

To extend the metaphor a little further, we need to be shopping for a flotilla—each boat smaller than the big vessel we've come to love, and each able to sail under its own steam but in the same waters with the rest.

Accomplishing this will take a combination of policies, markets, and technologies. All three will be necessary, much like a three-legged stool that will topple if one leg is missing. My own sense is that, given our current situation, forward movement will start with properly aligned policies, both public and private. The policy leg of the stool will need to be sturdier than the other two in the immediate future, in order to jump-start markets that will stimulate, support, and sustain further technological innovation and development.

Public Policy

The rules have to be changed. That is the role of the political process, in the best sense.

—Timothy J. Wirth

It's fashionable these days to associate policy with government and to disdain both the usefulness and appropriateness of government in the marketplace. We've seen examples of inappropriate government interference in the marketplace and the regrettable unintended consequences. The solar tax credit of the 1970s is one such instance.

Conversely, there are also examples of public policy jump-starting markets that would have never existed but for a little help from government. An example from history is the landmark New Deal legislation that created rural electric associations and succeeded in electrifying America's heartland during the mid-twentieth century. Today's examples include well-constructed state RPS policies. Lawmakers in Texas did an exceptionally good job of structuring the state's RPS (creating an effectual mix of meaningful incentives and penalties, for one thing), and they jump-started a market for utility-scale wind that likely wouldn't have developed without initial public policy intervention. Ideally, good policy leads to vibrant and self-sustaining markets. (Then we forget that policy was the first step, and we extol the "miracle" of the marketplace.)

Similarly, policies set by private-sector companies and individuals can have significant impact in the marketplace and can drive markets for new technologies. A handful of companies and corporations are undertaking some bold clean and green initiatives. Because they make economic sense while conveying an ethical message, these kinds of efforts are likely to catch on and spread throughout the corporate community. Needless to say, there also are examples of effective public-private partnerships.

Above all, whether we're talking public or private sector, there's one overarching policy approach that makes sense for all: no regrets. In brief, no regrets means there are multiple reasons to take action. Even if one reason turns out to be less compelling than originally thought, there are other reasons to act. Consequently, there are no regrets.

This reasoning has direct application to the climate change discussion. I don't believe there's any credible or scientifically sound basis on which to question the very real, overwhelmingly urgent climate crisis facing our planet, our carbon-addicted economy, and our heirs. Nevertheless, just for the sake of argument, if this issue were to prove less grave than we think, there are other compelling reasons to proactively, purposefully, and aggressively invest in energy efficiency and clean renewable energy alternatives. These reasons have to do with the economy, the environment, electric reliability and power quality, human health, homeland security, national security, intergenerational equity, resource depletion—all the drivers we have discussed.

Energy efficiency in particular is the mother of all no-regrets technologies, because the cheapest, cleanest, most-reliable energy is the energy we *don't* produce, distribute, store, and use. Amory Lovins coined the term *negawatt* to describe this negative megawatt.

We can not overstate the power of public policy to create markets for technologies. The Union of Concerned Scientists estimates that, if the states that have established an RPS actually meet their goals, they will add 50,000 MW of renewable power to the grid. This would be the equivalent of more than 140 small- or medium-sized power plants and it would result from a single public policy undertaken in fewer than half the states. That is the power of public policy.

Due to facilitating public policies, the largest market for PV these days is in Germany, a country situated at the same latitude as Anchorage, Alaska.* Similarly, one of the fastest growing markets for PV in recent years has been New Jersey. Chicago boasts PV on schools and government buildings. None of these places is warmed by a desert sun, but all enjoy a vibrant solar marketplace due to public policies.

Different kinds of policies are appropriate for different levels of government. For example, local and state governments are responsible for building codes. They can enact performance-based codes that include an energy operating budget. Being performance-based rather than prescriptive, these codes give builders and homeowners discretion in how they comply. This would favor energy efficiency, which is versatile and inexpensive. Like Aspen, code jurisdictions can create a financial penalty for energy hogs. This would create a disincentive for poor energy

* It's noteworthy that Germany, with a decidedly mediocre solar resource, captures a good part of the world's PV modules. One might wonder how much more solar electricity could be generated if the modules were directed toward parts of the world with a superior solar resource.

performance and a financial pool to support energy efficiency and distributed renewable energy products.

To upgrade the energy efficiency of the existing housing stock, local or state governments can require home sellers to make selected energy efficiency upgrades as a condition of resale. Added attic and wall insulation, water heater blankets, and other similar measures have been included in ordinances of several municipalities that took this approach in California and Wisconsin.

In their energy emergency planning, states and communities can ensure that operations critical to disaster mitigation, response, and recovery are powered by disaster-resistant, resilient electricity resources. Not only can PV generate needed critical electricity on-site without the need for fragile transmission and distribution wires. In addition, it's a meteorological fact that, after a hurricane passes, the sun shines brilliantly as part of the high-pressure system that follows the storm, thus providing added solar fuel for the system. Rooftop or building-integrated PV on police and fire stations, medical facilities, and gas stations should be incorporated in community disaster plans. Skid-mounted portable PV generating sets should be placed in National Guard armories and schools designated as emergency shelters throughout disaster-prone regions.

How many pictures did you see of the tragic situation at the New Orleans Superdome in which the sun was shining brilliantly on the floodwaters and human misery? Lives might have been saved and misery reduced if emergency responders and medical professionals had had electricity for critical needs in hospitals and nursing homes, and for communications capability in general.

During or following a disaster, a community's paramount electricity requirements are communications, life-support systems in medical facilities, emergency lighting, and fuel for emergency vehicles. Why is transportation fuel an electricity requirement? Because fuel pumps are powered by electricity. Following Hurricane Katrina, badly needed relief supplies were stalled in FEMA vehicles hundreds of miles north of the devastated Gulf Coast for lack of working pumps to fuel the vehicles as they traveled south.

Rooftop PV couldn't withstand the full-frontal assault of a category 4 hurricane or an F5 tornado. But PV panels can indeed weather severe storms, including hail bombardment.* One Colorado homeowner recalls that, after a particularly nasty hailstorm, he had to have his rooftop solar panels removed—but only to replace the damaged roofing shingles. The solar panels were fine, and he put them back on his new roof.

Regulation is a powerful public policy tool for incentivizing improved energy behavior. State governments regulate electric utilities, even in reregulated markets. Utility regulators must think outside the box and create meaningful incentives for utilities and other electricity providers to change the electricity generation paradigm in this country. Investment in energy efficiency and distributed generation should be encouraged and facilitated for all the reasons we have discussed.

In addition, government can lead by example, making sure its own buildings perform optimally. Any number of governmental entities, from the U.S. military to the small town of Carbondale, Colorado, have at least made the commitment on paper. Presidents, governors, and mayors have issued executive orders articulating "green" as a public policy goal. As an objective voice for the common

* NREL tests PV panels for weather tolerance if manufacturers request it. Some modules are left outside to weather the natural elements. Others are subjected to extreme heat and cold in indoor thermal chambers, where they also have been bombarded with homemade hail.

good, government can also inform citizens about new energy technologies, ways to conserve, and other relevant energy issues.

Government can make R & D investments in the public interest, much like the California Public Interest Energy Research program. Government can be an "investor of last resort," able to wait until last among creditors to reap the financial rewards, if any, of government-sponsored R & D. This is a role that's probably more appropriate to the federal government (and larger states) than local governments.

Leadership on electricity issues will likely come from states and communities. Just as all politics are local, so also are all actions. However, the federal government can exert powerful influence on our consumption of electricity by continuing to set performance standards for electricity-consuming equipment and buildings. As an example, some 2,000 builders have constructed more than 200,000 Energy Star homes, earning energy cost savings of some $60 million annually for occupants.*

Importantly, the federal government can set broad goals and a strategic framework for public policy. In the aftermath of the Arab Oil Embargo, Congress did a masterful job of articulating goals and creating a programmatic framework within which state and local governments could choose how to meet the goals.†

The Arab Oil Embargo created an obvious national emergency. Today, the emergency that confronts our country and the community of nations is global climate

* This is a voluntary government program in which, in return for use of a government agency's logo and informational support, builders agree to build to specified energy efficiency levels. Some states and cities have their own local programs as well.

† Initially, the feds also supplied needed funding. Starting in 1981 and thereafter, however, federal funding began to diminish substantially.

change.* Although this is likely the crisis of the ages, there has been a distressing dearth of discussion about what to *do* about it. Sunday news magazines have carried pieces pontificating glibly about "adaptation." While some degree of adaptation will be necessary in the coming years, it seems premature to default to this position before taking all feasible actions to *mitigate* the inexorable march of climate change.

It's the serious, measured discussion of feasible mitigation measures that has been lacking to date. The private sector has been as slow as the federal government to recognize and accept the phenomenon, much less adopt workable policies to help reduce the impact of this looming disaster.

The world's largest industry—insurance—is a case in point. The industry stands to suffer devastating financial consequences as the result of climate change impacts. Yet, it seems to be mired in collecting actuarial data and ascertaining the magnitude of the financial consequences for the industry. Insurers and reinsurers are examining the liability issues, yet it doesn't appear that they've addressed, in the industry's customary methodical and thorough fashion, the realm of actions and technologies that could mitigate the actual phenomenon of climate change. If the U.S. insurance industry were to do this, it could move entire markets for energy efficiency and renewables, because it's likely that the insurance industry would provide incentives for clean-energy investments by the insured.

Given the apparent inertia of both government and the private sector with regard to climate change, it's worth

* Any readers who are unconvinced that (1) global climate change is real, or (2) that the escalation in temperatures in recent decades is the consequence of human action, should read the following paragraphs about climate change with no-regrets policy in mind.

noting the usefulness of nongovernmental organizations (NGOs) in policy formulation. NGOs are groups of individuals who come together voluntarily in pursuit of common goals. One such NGO is the American Solar Energy Society (ASES). NGOs lack the bully pulpit and policy tools of government. Also, they aren't driven by the private sector's need to make a profit.

In 2006, under the leadership of former ASES chair Chuck Kutscher, ASES conducted a first-ever exercise. At its annual conference, it convened panels of experts in nine commercially available energy efficiency and renewable energy technologies.* Kutscher gave the experts common sets of assumptions and asked them to estimate the potential contribution of their technologies to the nation's energy mix by the year 2030.

The details are too many and too complicated to enumerate here. However, the exercise revealed that these technologies could achieve significant market penetration and, consequently, displace significant emissions of CO_2. The numbers were sufficient to put our country where it needs to be, on the ambitious path toward the dramatic CO_2 reductions that climate scientists believe will be necessary by mid-century.

These results may give us cause for a collective sigh of relief, but they're no occasion for glee. Essentially, ASES learned that, by utilizing commercially available technologies, we can tamp down CO_2 buildup so that it brings us merely to the brink of disaster but not past the tipping point. The huge value of this exercise, however, is that it's a signpost toward the road forward. ASES demonstrated,

* The technologies included energy efficiency in buildings and transportation; biofuels; the array of solar technologies, including utility-scale concentrating solar power; wind; and geothermal.

as no one else has to date, the potential power of policy, markets, and existing technologies to address our climate change emergency.[1]

One measure that will be called for is public policies that assign a value to carbon dioxide emissions. There is no national requirement to date. The most immediate and significant results could be achieved if emissions were taxed. Not surprisingly, the U.S. business community prefers a more market-based approach—creating markets for CO_2 offsets and actually trading carbon "credits." The effectiveness of this approach in reducing CO_2 emissions at the speed necessary to ward off disaster is questionable.

After years of resisting carbon regulation, some large corporations now ask for it. Big dogs such as Alcoa, Caterpillar, Duke Energy, and General Electric are calling on the federal government to regulate carbon in some manner. One infers that these companies would prefer to operate in one large carbon-constrained market (that perhaps they can help shape) rather than fifty little ones with different rules, conditions, and politics. Duke Energy's CEO has been quoted as saying, "If you're not at the table when these negotiations are going on, you're going to be on the menu."[2] Importantly, CO_2 regulation can be expected to boost the demand for technologies that aren't emitters. Although nuclear is one such technology, energy efficiency and renewable energy are considerably more affordable.

One wonders how much we could accomplish if we exerted the national will to make clean domestic energy the twenty-first-century version of a moon shot, such as Tom Friedman has called for. In any case, though, public policies shouldn't be *bi*partisan. They should be *non*partisan, just like policies developed in the private sector.

Corporate Policy

Increasing energy efficiency is not only good practice but it can also be good business.

—*Samuel W. Bodman, secretary of energy*

For a variety of reasons that make business sense to them and their shareholders, U.S. businesses are starting to go green. If the corporations are large enough, their purchases can drive entire markets. Some companies use *green*, *clean*, *sustainable*, and similar adjectives to distinguish themselves from the competition. Increasingly, some of them are adding negawatts to their power mix as a hedge against the future price, power quality, or reliability risks associated with conventional fuels. Some believe *green* or *sustainable* to be consistent with their corporate identity and product differentiation. Some may use *green* for image enhancement. Some may recognize their corporate role as influencers and try to lead by example. Some may use their unique bully pulpit for demonstration or public education purposes. No matter what the stated reason, you may be assured that a healthy corporate self-interest underlies these decisions. This is a good thing.

You won't find many bigger bully pulpits than Wal-Mart's big box stores. Whatever its corporate reasons, Wal-Mart has set the following goals for itself: become the United States' number one supplier of CFLs and sell 100 million of them in 2007; meet 100 percent of its building energy needs with renewable energy; and double the fuel efficiency of its trucks.[3] It's greening its big-box stores, starting with adding passive solar daylighting to some of them.* This is good

* An interesting, if unintended, consequence was that Wal-Mart noticed that merchandise displayed in the naturally day-lit sections of its stores sells better.

business. Energy is a billion-dollar cost *every year* for Wal-Mart, second only to labor. Wal-Mart calculates that it will save $52 million every year in fuel costs alone.

Now Wal-Mart is going solar more aggressively and has launched a ten-year pilot project to install PV at twenty-two of its stores in California and Hawaii. Each system could meet as much as 30 percent of the store's electricity needs. Wal-Mart calculates that this investment could offset greenhouse gases by 6,500 to 10,000 metric tons *every year*.

Astoundingly, however, this corporate effort might be derailed by public policy. It seems that the sunny island state of Hawaii, which must import all its fossil fuels at considerable cost, has enacted a statutory limitation on the size of solar arrays that may be placed on rooftops. The smaller array mandated by Hawaii would not meet Wal-Mart's criteria for the project. It's tragically ironic that, in this case, public policy might quash a corporate effort to "do the right thing." Moreover, for sunny Hawaii, where consumers pay around 30 cents per kWh for electricity, this would seem to be sadly counterproductive.

In addition, Wal-Mart has constructed two stores, one in Colorado and one in Texas, in which it's experimenting with a variety of sustainable technologies. These technologies include on-site wind generation, transpired solar collectors, PV, solar water heating, xeriscape (waterless) landscaping, and permeable asphalt pavements. Some of the technologies can be expected to help Wal-Mart manage its electricity costs in the future. Others appear to be efforts to be a good corporate citizen and steward of the local environment.

Recall our earlier discussion about the enormous electricity needs of IT-dependent businesses and the accompanying need for six (in the future, maybe even nine) nine's of reliability (99.9999 percent). Power quality is also a major concern with regard to sensitive electronic equipment. In an effort to hedge against these risks, Google will supplement its electricity with on-site solar electricity generation at its corporate headquarters. It recently announced a partnership with Sharp, a leading PV manufacturer, to provide the modules and install the largest commercial PV system in the United States.

In addition, Google plans to install two carports, the roofs of which will be composed of PV arrays. In a separate partnership with PG&E, Google will add hybrid electric cars to its fleet, converting them to plug-in hybrids and using the PV-powered carport to recharge the vehicle batteries. Unneeded solar electrons will be sold back to the grid. This adds a hip, futuristic spin that's consistent with Google's corporate image. It also provides critically important electricity backup to ensure reliability and adequate power quality. Because of Google's special status with younger Americans, perhaps its energy leadership will educate and inspire the coming generation of energy consumers.

As an electric utility, Chicago-based Exelon is an influencer and, it could be argued, has a special obligation to lead by example. That's what it's doing. Its new corporate headquarters meet U.S. Green Building Council standards.* Exelon calculates that it incurred a 5 percent premium in this major renovation. However, the energy-related measures are expected to pay for themselves in reduced energy consumption within five years. Thereafter, the cost savings

* USGBC instituted the LEED (Leadership in Energy and Environmental Design) standards. Points to attain different levels of LEED certification (platinum, gold, silver, or bronze) are awarded on a number of measures of sustainability. Energy is among the criteria but is by no means pivotal. Consequently, it's possible to build a LEED-certified building that doesn't perform particularly well from an energy perspective.

will inure to Exelon. This is good business and good corporate strategy, and it comes from a utility expert.

Among Exelon's sustainable practices was purchasing more than 60 percent of the construction materials from manufacturers located within a 500-mile radius. This avoided the emissions associated with transporting the goods. In addition, Exelon recycled 75 percent of the construction waste, and almost one-third of furniture and other materials were reused in order to reduce waste. Exelon is purchasing RECs from regional wind power facilities in order to offset 100 percent of its electricity use. Finally, 96 percent of its energy-using equipment—from copiers to motion sensors to lighting—is Energy Star certified.

Government and corporate influencers may lead by example, but Americans may be more likely to follow something that is nearer and dearer to their hearts: the home team. In the bully pulpit of their home field, the San Francisco Giants are partnering with the utility home team, PG&E, to install enough solar panels to light the scoreboard for the entire season. The Colorado Rockies also use the power of the sun to display the score.

Whole Foods believes that its image as a leader in natural and organic foods entails the obligation of stewardship for the environment in the communities where it's a corporate citizen. Consequently, Whole Foods' policy is to offset 100 percent of its electricity use with clean and renewable energy. In 2005, Whole Foods made what was at the time the largest-ever purchase of wind RECs in the United States.[4] The company calculated that this avoided more than 700 million pounds of CO_2 pollution—the equivalent of taking more than 60,000 cars off the road.

The impact of corporate decisions to operate in a sustainable or energy efficient manner can be multiplied several times over when companies require their downstream contractors to do the same. More companies are starting to flex their corporate purchasing power in this fashion.

Corporations create markets for energy efficiency and renewable energy technologies by looking out for their own self-interest. This is particularly true if their bottom line could be directly affected by climate change. Colorado ski resorts, for example, recognize that climate change can shut down their high-speed lifts. Predictions of shorter ski seasons and reduced snowpack translate into a four-letter word: risk.

Consequently, Aspen Ski Company filed an amicus brief with the Supreme Court when several parties challenged EPA's assertion that it lacked authority to regulate carbon dioxide. The petitioners prevailed in a landmark ruling that's expected to help pave the way to CO_2 regulation. Aspen also pursues a number of energy efficiency and renewable energy improvements in its own operations. Vail Resorts, one of Aspen's competitors for destination skiers, purchases Colorado wind RECs to offset 100 percent of its electrical load.

Not only can corporations create and drive new markets for energy efficiency and renewable energy through their internal policies (purchasing practices being prime among them), they can also drive public policy. At a time when our national government often seems paralyzed by competing interests and conflicting policy goals, nonpartisan no-regrets energy policies are increasingly prevalent in the private sector and may be among the bright spots on the energy horizon.

Markets

We have a multigenerational problem that requires a systemic, multigenerational response, and that can happen only if we get our energy prices right. Only that will guarantee green innovation and commercialization at scale.

—Thomas L. Friedman

Shell bought me a root canal.

—Eddie Rhoderick, rancher, Briscoe County, Texas

Both statements articulate the situation today and in the near future with regard to markets for clean-energy technologies. Friedman understands the big picture, the macro view: in order to stimulate the growth of emerging clean technologies, we need to get energy prices right. That is, we need to realign policy incentives to stimulate behaviors that meet today's needs. We also need to curtail subsidies and other financial inducements producing results that, in today's changing world, no longer meet our societal needs and have actually become counterproductive. I would add that we need to cut off the free rides for industries that require society to bear the costs of their harmful and expensive externalities.

In contrast, Eddie Rhoderick observes the impact on the ground, the micro view, from his ranch on the windy Texas plains. Shell has leased 1,900 acres of his land for development of a wind farm. The lease is lucrative enough to avail Rhoderick of an expensive root canal.

As I've said before, there's no such thing as a free market for energy and there never has been. Now, however, new markets for renewable energy and energy efficiency are starting to take off, even though some of us might have already forgotten that some of them were jump-started by public policy. As a general proposition, these markets will be characterized and driven by several factors: (1) new financial instruments and tools, (2) new partnerships, and (3) new approaches to risk.

New Financial Instruments and Tools

We've talked about renewable energy credits (RECs), also known as green tags. These are financial instruments that reflect the environmental attributes of renewable energy technologies. Customarily, one of these tradable units is the equivalent of the environmental attributes of one megawatt-hour of electricity from a renewable generation source. Anyone wishing to help stimulate the growth of markets for renewable energy technologies can contribute to reducing the technologies' cost—and boosting their market presence—by helping to capitalize them through the purchase of green tags. This also works for companies that need to comply with a renewable energy mandate of some kind.

A similar instrument is energy savings certificates, also known as "white tags." As the name indicates, white tags will do for energy efficiency what green tags do for renewables. This market is brand-new and has not taken off just yet, but it will operate as the REC market does. An authorized body will issue the tags, guaranteeing that a specified amount of energy savings has been achieved. Each certificate will be a unique and verifiable commodity, guaranteeing that the savings haven't been achieved elsewhere.

The last is a critical point and applies to both white and green tags, as well as carbon credits. For them to have practical value, intellectual honesty, and credibility in the marketplace, each certificate must represent units of value that aren't being counted elsewhere. This is the issue of additionality.

Carbon credits are similar, except they represent offsets and will be used for environmental compliance. Unless a carbon tax is instituted, it's likely that climate change regulation will be addressed through the instrument of carbon credits—bought and traded in authorized exchanges, and recorded to document compliance. This is commonly called cap-and-trade. Typically, the government sets pollution limits (caps) and then auctions or allocates shares of credits among industries. This kind of trading scheme has worked with sulfur dioxide.

It's important that government get this right or the credits will not reflect a meaningful market price. This happened in the European Union. When the EU allocated too many credits, the market was flooded and the credits were devalued.

Those who are unable to comply with CO_2 caps would purchase credits from those who have them to sell. One venue for this market has already been created: the Chicago Climate Exchange. This voluntary exchange has been growing since its inauguration four years ago and now numbers 225 members. Carbon financial instruments (CFIs) are registered under the Exchange and are traded among member companies. Utility experts estimate that the market price will need to rise to $30 per ton of CO_2 over time in order to achieve the desired results.

Regional markets for carbon credits are already growing as the result of public policy. Nine northeastern states banded together to form the Regional Greenhouse Gas Initiative and are scheduled to start trading carbon credits amongst themselves in 2009. California will be a major player in the West, having recently enacted greenhouse gas limits. It also created the California Climate Action Registry, a voluntary program under which companies and organizations may register an inventory of their greenhouse gases. This will establish a baseline against which future reductions can be measured.

Tradable carbon credits, RECs, and white tags open the renewable energy marketplace to participation by a whole new set of players. Some of them literally *are* players—football players. Take the 2007 Super Bowl in Miami. This Florida city has no wind resource to speak of, so the NFL purchased RECs to support wind elsewhere. The amount of the purchase was calculated to offset the 500 tons of CO_2 emissions resulting from additional electricity requirements and traffic generated by the game. In addition, the NFL planted trees in South Florida to eventually recapture the CO_2 added by the event.

New financial tools and approaches are being created to address the capital cost issue associated with renewable energy technologies. Third-party financing is emerging as an affordable option for small businesses and commercial operations wishing to buy solar. Now they may purchase solar *electricity* without the hassle and up-front cost of buying the hardware. Third-party financiers fund, install, operate, maintain, and—importantly—*own* the solar energy systems. The building owner may lease roof space to the third party and purchase the electricity generated at the installation.

One such arrangement in Boulder, Colorado, will permit a community hospital to lease rooftop space to a solar installer in exchange for a flat electricity rate from the 55 kW system over a ten-year period. At the end of the contract, the hospital may choose to purchase the system. In the meantime, it will enjoy lower and more-predictable electricity costs in exchange for transferring the RECs associated with the project to the solar company.

This is a win-win deal. The hospital enjoys the ease of a turnkey system, because the solar company will operate and maintain it. From a business perspective, the hospital also enjoys lower and more predictable energy costs. The solar company will get to claim the value of the RECs from the electric utility, which is mandated under the state's RPS to generate 20 percent of its electricity from renewable energy by the year 2020.

Lenders seeking a market niche now bill themselves as green. They make loans for both energy efficiency and renewables, and they underwrite green mortgages as well. Energy efficiency loans can pay for retrofits such as added insulation, efficient water heaters, and other equipment that increases the energy efficiency of the structure. These loans are like any other, except the lender can factor in monthly savings on utility bills and credit the debt accordingly, permitting potential borrowers to qualify for larger loans than they might otherwise be able to secure.

New Partnerships

Interesting partnerships are being formed to strike innovative and lucrative deals in the emerging clean-electricity marketplace. Particularly interesting is a deal made in Texas in which investment banker Goldman-Sachs assisted two private equity firms in purchasing Texas Utilities (TXU), a coal-fired utility giant. TXU had announced plans to construct eleven new coal-fired power plants in the Lone Star State, and environmental groups were up in arms. Goldman-Sachs brokered negotiations between the potential purchasers and the environmental groups. The upshot was that the new buyers agreed to scrap plans to build all but three of the planned coal plants, agreeing to serve projected load growth through efficiency and renewables.

So far, this looks like a win-win compromise for everyone. The investors win, because they need to invest their capital in timely fashion and now they have a deal.

The environmentalists win, because eight coal plants will be taken off the planning books. The investment banker wins, because the deal was made. Moreover, it appears that the risk of future costly litigation has been averted with the environmentalists' agreement not to oppose the three remaining coal plants.

Two other interesting partnerships are coalescing around lighting issues. About 22 percent of our electricity powers lighting, and 42 percent of that (9 percent of all electricity) currently powers incandescent lights (of which, recall, 90 percent is thrown off as waste heat). If improved efficiency could halve that figure, it has been estimated that it would equal two or three years' growth in electric demand.[5]

Consequently, utility giant Duke Energy (which generates mostly with coal), the Alliance to Save Energy, and several lightbulb manufacturers have banded together to persuade consumers to change out the bulbs in the United States' four billion sockets. This coalition encourages consumers to purchase CFLs and other energy efficient lighting technologies, and urges a "market phaseout" of incandescents by the year 2016. They calculate that a phaseout could save $18 billion every year in electricity. The amount of power saved would equate to the output of thirty nuclear reactors or as many as eighty coal plants (with the associated savings in externalities for both).

In contrast, another coalition that includes General Electric (a company whose roots extend to Thomas Edison, the inventor of the incandescent lightbulb) and the Natural Resources Defense Council (a powerful environmental group) argue against outlawing incandescent bulbs. This group advocates improving the energy efficiency of incandescent bulbs.

A California utility is partnering with IBM, utilizing

its technology to make data centers more energy efficient. Data centers can use up to 100 times the energy, on a square-foot basis, of typical office space. PG&E is leading a coalition that includes utilities in all regions of the country, as those regions experience huge new demands for electricity infrastructure to meet the growing demands of data centers. Data center operators are constructing new facilities, creating burgeoning new electricity requirements both through the IT operations themselves and the thermal space conditioning required by the delicate machines.

Communities themselves are getting into the marketplace as participants in collaboratives that own renewable energy projects. Local owners might be citizens, much like any other rural cooperative, who share the profits of selling renewably generated electricity to the grid. Or the collaborative might be a public-private partnership in which the private partners can benefit from the tax advantages and the public partners can contribute their ability to obtain inexpensive financing. In any case, partners are chosen based on the complementary financial attributes they bring to the partnership.

One example is the "flip" model, in which ownership initially resides with an entity that can benefit from the tax credits but eventually flips back to the local owners after the tax benefits are exhausted. Farmers in Minnesota and Iowa, residing as they do in the region dubbed "the Saudi Arabia of wind," have been pioneers in these kinds of creative financing arrangements, following models created in Europe, particularly Denmark. One of the first U.S. models was MinWind in Minnesota. When started in 2000, the project consisted of four 950-kW turbines, owned by sixty-six local farmers. Seven more turbines were added in 2004. This second group is owned by some 200 local investors.

The operative word here is *local.* There are numerous

benefits of local ownership. Opposition to utility-scale developments is minimized by comparison to out-of-area, developer-owned projects. If the electricity is consumed locally, the need for large transmission towers and lines is minimized. Importantly, the economic multiplier kicks in, yielding $4 million annually in local income—more than three times the local income generated by traditional developer arrangements.[6]

No matter what kind of partnership is formed, much of the investment in distributed renewable energy technologies in future years will be shaped by the art of the deal. Utility ownership of PV arrays on privately owned rooftops, flip model ownership of community wind turbines, and so forth; whatever the arrangement, if properly designed and targeted financial incentives are available, deals can be struck to underwrite the daunting first costs of renewables. This can be expected to grow the market for these technologies and thus contribute to their eventual affordability.

New Approaches to Risk

Risk-hedging may be what's driving Shell Oil to diversify its energy portfolio to include wind energy. No doubt Shell has a number of reasons for wanting to expand into different markets. The risks of peak oil, nationalization of overseas oil fields, and possible CO_2 regulation might be among them.

Key to future win-win deals will be Wall Street's understanding of energy efficiency and renewable energy technologies, as well as a new appreciation of the *real* risks posed by these technologies, compared to twentieth-century conventional technologies. Trying to accelerate the pace of Wall Street's progress up this learning curve, an interesting coalition of NGOs and private sector firms has emerged.

Called Ceres, it's a coalition of more than eighty investor, environmental, and public interest organizations. Ceres' goal is to educate the financial community about energy efficiency and renewable energy technologies. In particular, Ceres aims to improve investors' and companies' research on the financial risks and opportunities associated with climate change and it calls for climate risk to be a consideration in standard financial analysis. It also works with investors and the Securities and Exchange Commission to improve corporate disclosure about climate change exposure and potential liability. Ceres' work will become even more important if carbon emissions are regulated in the future.

Among the industries that can be expected to encourage the growth of energy efficiency and renewable energy for risk-hedging purposes is the insurance industry. Losses from weather-related events have topped half a trillion dollars over the past twenty-five years, and the U.S. General Accountability Office has issued a report warning of more climate change–induced extreme weather to come. This could spell doom (another four-letter word) for U.S. insurance companies. Inasmuch as insurance comprises 10 percent of the U.S. economy, the impacts would ripple throughout all markets. It's only a matter of time before insurers, just like European reinsurers, will require the insured to adopt climate change policies and certify that they're undertaking mitigation actions. One would expect energy efficiency to be prime among the actions, followed closely by renewable energy.

Just as our model of electricity generation and distribution will look very different in the coming years, so too will the financial tools and approaches that underwrite these systems. These financial models will capitalize on the tools and opportunities created by policy.

Technologies

Invention is one percent inspiration and 99 percent perspiration.

—*Thomas Edison*

Here's the inspiration: picture a plug-in hybrid or fuel-cell vehicle. It looks like other vehicles, except it runs on both petroleum (or biofuels) and electricity. When you get home in the evening, you plug your vehicle into your house to recharge your battery or fuel cell. Sometimes you might decide to reverse the flow of electrons and power your house off your vehicle.

Your house is oriented toward the sun. It has building-integrated PV arrays that cover the roof and perhaps part of a south-facing wall. You might have your own wind turbine. Perhaps your neighborhood has a community-scale wind turbine or a district heating and cooling system. You might have a geothermal heat pump. There's an electrolyzer in the utility room (powered by your solar or wind equipment) to split molecules of water into hydrogen and oxygen, and you have a small fuel cell to convert the hydrogen to electricity. Let's put a hydrogen storage tank underground, as well, to store the electricity your house makes.

You're grid-connected. Your enlightened utility has a net metering policy, and you sell excess electricity back to the utility. Your home computer gives you access to a cyberspace marketplace for electricity. You buy and sell electricity in real time. Your regulators permit time-of-day pricing, so you can schedule your electricity use to minimize costs and you can try to sell your electrons at peak times to earn top dollar for them. If your neighbors have the same electricity generating and storage capability you do, you can create your own neighborhood microgrid.

All in all, you inhabit a net- or near-zero energy home, you fuel your own vehicles, and you might even sell electricity and make a little money on the side.

Futuristic? Yes. But it's not that far out—thirty years, maybe a little sooner.

That was the inspiration. Here's the perspiration: many needed technology developments separate our electricity system today from this exciting vision of tomorrow. First and foremost, we need to develop energy-storage devices for intermittent renewables. Batteries are one technology and hydrogen is another, especially for utility-scale generation.

From the individual PV owner's perspective, backup with today's batteries doubles the price of a PV system. In addition, today's lead acid batteries need to be replaced several times over the life of the PV system, and then they need to be responsibly disposed of. This is expensive, it's a hassle, and it creates a maintenance chore that otherwise would not exist. Most owners of residential PV systems eschew battery backup and rely instead on the utility's grid. The consequence is that, if the grid goes down for some reason, so does your PV. If you install PV for reliability purposes, you must invest in battery backup.

Improved batteries will be lighter weight and longer lasting than today's batteries. This will benefit electric and hybrid electric vehicles. In addition, improved batteries will facilitate the transition from hybrid vehicles to plug-in hybrids (PHEVs). PHEVs, and the suite of technologies that enables them, will help pave the way toward our house of the future. They also will revolutionize our electricity system in two ways: (1) they'll enable utilities to sell electricity at night, when drivers recharge their vehicles at home, thus filling in those overnight valleys and helping utilities achieve their ideal load shape of a relatively flat horizontal line, and (2) they'll blur the traditional lines separating the electricity, buildings, and transportation sectors, especially if electrons flow back into the electricity grid from the PHEVs at peak demand times.

From the utility's perspective, a key limitation of solar and wind technologies is that the electricity isn't dispatchable. That is, the electrons can't be stored for use by the utility when needed. It turns out, though, that electricity can be made from hydrogen, and hydrogen can be made by electrolyzing water. If the electric charge to do that is supplied by renewable energy, the sun and wind can be used to make the very storage medium for their power. Lack of dispatchability is the key limitation of utility-scale electricity generated from solar and wind. Storage would reduce or eliminate this limitation.

Smart meters and load-control devices will help defer or avoid investment in expensive new power plants. Smart meters will be your "phone line" to and from your utility. Assuming that your regulators allow time-of-day pricing, your smart meter will inform you about the price of electricity throughout the day. You can reschedule much of your electricity using equipment for off-peak, cheaper times—just like scheduling long-distance or cell phone calls depending on your plan. If a lot of people do this, our utility "chef" may start to fill in some valleys at the Demand Café. This technology will kick in long before the home of the future and will be among the suite of technologies that will help get us there.

Besides providing real-time price signals to consumers, smart meters will also make utilities more knowledgeable about consumption patterns on their system. This will facilitate their efforts to "mine" the untapped energy efficiency potential before building new power plants.

Perhaps you permit your utility to cut back your electricity use whenever it needs additional power. In return, it pays you or puts you on a cheaper rate. The utility uses a load-control device that it runs remotely. These devices are actually in limited use today, permitting utilities to manage electricity consumption at peak demand times. Customarily, they reduce your load for just a few minutes. You hardly notice the loss, if at all. But by aggregating these devices at thousands of meters across its service area, a utility can cut back enough to avoid purchasing expensive peak power or suffering rolling blackouts.

This brings us to the concept of a smart energy network. Hailed by Pacific Northwest National Laboratories as "electricity's third great revolution"[7] (after Edison's incandescent lightbulb and Tesla's alternating current), the smart network will use cheap computing power and low-cost bandwidth to monitor and optimize transmission grid inputs and operations. It will schedule and integrate distributed generation systems into the grid, and it will help the system recover after disasters and other "events" that disrupt power. It will use twenty-first-century telecommunications capability to enhance power operations and delivery.

Is there a third generation of PV technology waiting to be developed? We don't know for sure, but scientists are working on it. Recall that the first generation is today's commercially available silicon-based technology. The second generation, already in the marketplace but in need of significant further development, are the thin films. The third generation could be nanosolar—quantum dots. The efficiency and cost-effectiveness could revolutionize the field of solar electricity—boosting the conversion efficiency to as high as 60 percent—but it's too soon to tell if the concept will pan out.

Much of the future of utility-scale wind lies offshore. The East Coast continental shelf, as well as the Texas Gulf

Coast, provide mooring for tomorrow's huge 6.5 MW (or larger) turbines that are too big to transport inland. Inasmuch as the best winds are found offshore, and the twenty-eight states with oceanfront property constitute 78 percent of our national utility load, these marine wind farms could contribute significantly to DOE's goal of supplying 20 percent of U.S. electricity needs from wind by the year 2020.

Complementing the R & D in big offshore wind is DOE's parallel effort to capture the energy of low-speed winds. This would be particularly useful for urban turbines and community-scale wind development. Moreover, it would allow us to mine much more plentiful veins of wind energy throughout the country. There are twenty times more low-speed wind sites on the Great Plains than there are high-speed sites considered appropriate for today's utility scale turbines.

As for our homes and businesses, net- or near-zero energy buildings, constructed from aggressive energy efficiency and distributed on-site renewables, will become more prevalent. In certain situations, community-scale net- or near-zero energy may be possible. This would entail aggressive energy efficiency, community-scale wind turbines, district heating and cooling systems, and the like. The financial models to underwrite the cost already exist, in part courtesy of those pioneering farmers on the Great Plains.

The future isn't just about developing new technologies. It's also about improving manufacturing technologies to mass produce today's (and tomorrow's) technologies more cost efficiently. And it's about finding new applications and new customers for today's energy efficiency and renewable energy technologies. Already, Habitat for Humanity and a few affordable-housing developers are starting to design passive solar buildings and equip them with energy efficient appliances and rooftop solar.

Technology developments are coming at us so fast that we can't keep up. I've been stockpiling reports and articles while researching this book, and I'm astounded at what's out there. Superconducting wires for bulk power transmission can reduce the large line losses that contribute a large part of our collective electricity waste. The conversion of turkey and other poultry litter to electricity generating fuel already is happening in Minnesota. Which technologies will pan out and which ones won't … who knows? The one thing I do know for sure is that I haven't *begun* to tell you about all the technologies and innovations detailed in the articles in my stack.

Technology-wise, the future is indeed so bright that we have to wear shades.

Conclusion

> *If we don't change the direction we are headed, we will end up where we are going.*
>
> —*Chinese Proverb*

Recall that there are no silver bullets in our energy arsenal. Fortunately, however, there is plenty of silver, maybe even platinum, buckshot. The buckshot is not limited to technology, but also includes innovative policy and financial ammunition. If we as a country are agreed on our energy goals, we can get ready, aim, and start firing at the energy bull's-eye.

Importantly, we must examine our current energy situation and understand where it's taking us. If we don't like the direction in which we're headed—resource depletion, rising fuel costs, massive long-lived environmental and health damage, to name a few—then we need to change where we're going. This will require changing what we're *doing*.

I'm not talking about an overnight energy revolution. Really, it's more like an *evolution*, incorporating both twentieth- and twenty-first-century technologies as we transition to the completely different, carbon-constrained energy reality in the coming years. For the reasons we've discussed, twenty-first-century renewable energy and energy efficiency technologies are in the wings and have walk-on roles, but they're not ready to play the lead. In the coming years, they'll play bigger and bigger supporting roles, as the aging twentieth-century electricity stars retire. Ultimately, they will progress to the limelight (along with technologies yet to be invented).

The word *revolution* is too abrupt and scary, but *evolution* is too leisurely. I'm thinking of what needs to happen as *urgent evolution*. We need a better word. But whatever we call this urgent evolution, our attitude toward it should be welcoming, not fearful. Think of the excitement with which we greet the latest innovations in cell phone technology, yet how many consumers even heard of cell phones a generation ago? Think of how exciting the new computer and TV technologies are. Who ever heard of plasma TV, yet now it's advertised as the perfect gift for dad on Father's Day. We should feel that same excitement about changing the electricity systems that power those new TVs and the computers that enable us to shop online to get them.

Renewable energy and energy efficiency can be expected to develop a larger presence in the marketplace, especially as the cost of twentieth-century fuels continues to go up and the capital costs of renewables continue to go down. Energy efficiency in particular will gain significantly greater market share because of its no-regrets nature and the fact that it doesn't require the investment of materials needed by utility-scale technologies. Twentieth- and twenty-first-century utility technologies alike will require significant amounts of steel, cement, and other commodities that will be in increasingly short supply globally and will bear increasingly hefty price tags as a consequence.

We will continue to rely heavily on central station power plants, especially in the near- and mid-term, because our electricity needs are humongous and continue to grow. Twenty or thirty years from now, however, the home of the future that was described a few pages ago could become a reality.

In the meantime, there's a critically important factor of our electricity equation that I haven't talked much about: You. The old adage "think globally, act locally" was never truer. There are several websites that tell you how to calculate your footprint. You can conserve electricity through your own actions. You can purchase energy efficient lightbulbs. You can start to use canvas bags for your smaller shopping trips, reducing the carbon footprint associated with petroleum-based plastic bags. You can question your utility decision makers about your utility's generating fuels. You can urge your legislators to enact laws that encourage the development of clean energy. You can make efforts to limit your own carbon footprint.

Above all, you can be responsible for your own actions. I quote Jerry Garcia (tongue in cheek, of course): "Somebody has to do something, and it's just incredibly pathetic that it has to be us." A more positive spin on the same message comes from renowned anthropologist Margaret Mead: "Never doubt that a small group of committed citizens can change the world. Indeed, it is the only thing that ever has."

We as individuals, as neighbors, as citizens of our

towns and states, can lead our government. In fact, to date, the most effective clean energy and CO_2-limiting initiatives have emanated from the local and state realms. Our roots as a nation are in the grass. We know how to do this. We can lead our leaders in Washington. As Winston Churchill said, "Americans usually do the right thing, but only after exhausting all the other options."

Carpe ventem: catch the wind.

Notes

Introduction

1. Richard Munson, *From Edison to Enron: The Business of Power and What It Means for the Future of Electricity* (Westport, CT: Praeger Publishers, 2005), 20.
2. Joel Makower, Ron Pernick, and Clint Wilder, *Clean Energy Trends 2007* (San Francisco and Portland: Clean Edge, Inc., March 2007), 2, www.cleanedge.com/reports/Trends2007.pdf.

Energy Schmenergy

1. Senate committee on Foreign Relations, *U.S.-India Atomic Energy Cooperation: The Indian Separation Plan and Administration's Legislative Proposal*, 109th Cong., 2nd sess., April 5, 2006.
2. McFarlane, Robert, "Colorado New Energy Summit 2007," (keynote address, Colorado New Energy Summit 2007, Denver, March 24, 2007).
3. Aldo Svaldi and Zach Fox, "The Incredible Edible Expense," *The Denver Post*, July 10, 2007.
4. Ricardo Bayon, "The Fuel Subsidy We Need," *The Atlantic Monthly*, January/February 2003, 117–118.
5. Jeffrey Kluger, "Global Warming Heats Up," *Time*, April 3, 2006.
6. Natural Resources Defense Council, *A Responsible Energy Plan for America*, report, 2005.
7. Ken Silverstein, "Mercury Debate Reaches Fever Pitch," *EnergyBiz Insider*, September 16, 2005, www.energycentral

.com/site/newsletters/ebi.cfm?id=28.

8. Diana Farrell, Scott S. Nyquist, and Matthew C. Rogers, "Making the Most of the World's Energy Resources," *The McKinsey Quarterly*, no. 1, 2007.
9. "Oil's Up to Stay; Start Buying," *The Denver Post*, February 24, 2007.
10. Whipple, Tom, "Peak Oil Review," *Association for the Study of Peak Oil and Gas-U.S.A.*, March 5, 2007.

Electricity 101

1. Ken Silverstein, "Solar Getting Limelight," *EnergyBiz Insider*, March 19, 2007, www.energycentral.com/centers/energybiz/ebi_detail.cfm?id=295.
2. Silverstein, "Limelight."
3. Energy Watch Group, "Coal: Resources and Future Production," March 2007, 6.
4. Richard Munson, *From Edison to Enron: The Business of Power and What It Means for the Future of Electricity* (Westport, CT: Praeger Publishers, 2005), 98.
5. Silverstein, "Limelight."
6. Hydropower costs around 2 cents, natural gas about 8, new coal 5 or more (depending on rail hauling rates and whether a carbon penalty is enacted), energy efficiency at 2 to 3 cents, and wind at around 5 cents. All costs are approximations. Not only does cost vary by technology, it also differs among utilities and it changes over time. The cost of construction for any technology that uses steel or cement, for example, has risen significantly in recent years due to world demand for these commodities. Silverstein, "Limelight."
7. Munson, *Edison to Enron*, 81.
8. Elizabeth Thomson, "Lack of Fuel May Limit U.S. Nuclear Power Expansion," MIT Tech Talk, April 4, 2007.
9. James L. Conca and Mick Apted, "Nuclear Energy and Radioactive Waste Disposal in the Age of Recycling" (paper

presented at Global 2007: Advanced Nuclear Fuel Cycles and Systems, Boise, ID, September 9–13, 2007).

10. John Carey, "Coal: Could Be the End of the Line," *BusinessWeek*, November 13, 2006, www.businessweek.com/magazine/content/06_46/b4009089.htm?chan=search.
11. Hermann Scheer, *Energy Autonomy: The Economic, Social, and Technological Case for Renewable Energy* (London: Earthscan Publications, 2007), 121.

Electricity 2000+

1. The Edison Foundation. 2006. "Why are electricity prices increasing? An industry-wide perspective," As quoted in Ceres. 2007. *TXU's Expansion Proposal: A Risk for Investors.*
2. Kevin Monroe, *TWP Green Sheet*, March 7, 2007, quoting Dr. Dan Arvizu, director, National Renewable Energy Laboratory.
3. National Renewable Energy Laboratory, "Buildings," *Fact Sheet* October 2006.
4. B. C. Farhar and T. C. Coburn, *A New Market Paradigm for Zero-Energy Homes: The Comparative San Diego Case Study*, NREL Technical Report (NREL/TP-550-38304-01). December 2006.
5. Bob Fesmire, "Is Efficiency the New Reliability?" *EnergyBiz Insider*, April 16, 2007.
6. Robert A. Hawsey, presentation, June 19, 2007.
7. U.S. Department of Energy, *Solar America Initiative: A Plan for the Integrated Research, Development, and Market Transformation of Solar Energy Technologies*, February 5, 2007, 5. Available at www1.eere.energy.gov/solar.

What Are We Waiting For?

1. Dr. Lawrence M. Murphy, phone interview, July 2, 2007.
2. Jim Hock, "TechNet CEOs Unveil 'Green Tech' Policy Agenda," *Press Release*, March 14, 2007.
3. Hermann Scheer, *Energy Autonomy: The Economic, Social, and*

Technological Case for Renewable Energy (London: Earthscan Publications, 2007), 131.

4. Scheer, *Energy Autonomy*, 121.

The Future

1. The ASES report is titled *Tackling Climate Change in the U.S.: Potential Carbon Emissions Reductions from Energy Efficiency and Renewable Energy by 2030.* It may be downloaded from the ASES website at www.ases.org.
2. Joel Makower, Ron Pernick, and Clint Wilder, *Clean Energy Trends 2007* (San Francisco and Portland: Clean Edge, Inc., March 2007), 7, www.cleanedge.com/reports/Trends2007.pdf.
3. Ibid., 13.
4. Whole Foods Market, "Whole Foods Market Makes Largest Ever Purchase of Wind Energy Credits in United States," press release, January 10, 2006.
5. Matthew L. Wald, "A U.S. Alliance to Update the Light Bulb," *The New York Times*, March 13, 2007.
6. Greg Pahl, "Community Supported Energy Offers a Third Way," March 12, 2007, www.renewableenergyaccess.com.
7. Patrick Mazza. "The Smart Energy Network: Electricity's Third Great Revolution." A report from Climate Solutions, Olympia, WA (based on an in-depth paper, available at www.climate solutions.org).

Resources

For information on energy efficiency and renewable energy technologies:

www.nrel.gov

The website of the National Renewable Energy Laboratory. A vast resource, providing one-stop shopping both through the volume of information on the site and the many links to other information. This site includes everything from "what is it and how does it work" information geared toward the elementary school level, to sophisticated technological materials, to a digital library of thousands of photos—all downloadable and free.

www.eere.energy.doe.gov

Like the NREL website, a rich source of information on renewable and energy efficiency technologies. Also includes information on facilitating policies and programs in the federal government, as well as state by state.

For information on energy efficiency:

www.aceee.org

The website of the American Council for an Energy Efficient Economy. Academic reports and white papers about policy issues. Brochures and reports rating consumer products.

www.ase.org
The Alliance to Save Energy. Information on energy efficiency technologies, products, and policy actions.

For information on state renewable energy and energy efficiency policies:

www.dsireusa.org
The Database of State Incentives for Renewable Energy. It's updated quarterly and contains information on state legislation and rules with regard to energy efficiency and renewable energy.

www.naseo.org
The website of the National Association of State Energy Officials (directors of state energy offices). It contains reports and information on a variety of energy policy issues.

www.ncsl.org
The website of the National Conference of State Legislatures. Includes downloadable policy reports as well as information about state legislative initiatives.

For information on renewable energy products near you:

Check out the website of the solar energy industries association in your state. If you don't know what it's called, check the national website, www.seia.org.

For information on climate change and how to calculate your carbon footprint:

www.epa.gov (Click on "climate change.")

About the Author

Carol Sue Tombari currently works at the U.S. Department of Energy's National Renewable Energy Laboratory and has specialized in energy and environmental policy and

programs for more than twenty-five years. She directed the State of Texas's energy efficiency and renewable energy programs for a decade, served as natural resources advisor to the lieutenant governor, and helped found the National Association of State Energy Officials. In addition, she was appointed to federal advisory posts by two secretaries of energy—chairing a congressional advisory committee on the subject of renewable energy joint ventures and serving on DOE's State Energy Advisory Board. Ultimately, it's her love for the next generation that continues to drive her work to protect the future of our planet and the lives of those yet to come.